Work, Race, and the Emergence of Radical Right Corporatism in Imperial Germany

Social History, Popular Culture, and Politics in Germany
Geoff Eley, Series Editor

Recent Titles

After the Nazi Racial State: Difference and Democracy in Germany and Europe, Rita Chin, Heide Fehrenbach, Geoff Eley, and Atina Grossmann
Work, Race, and the Emergence of Radical Right Corporatism in Imperial Germany, Dennis Sweeney
The German Patient: Crisis and Recovery in Postwar Culture Jennifer M. Kapczynski
Topographies of Class: Modern Architecture and Mass Society in Weimar Berlin, Sabine Hake
Neither German nor Pole: Catholicism and National Indifference in a Central European Borderland, James E. Bjork
Beyond Berlin: Twelve German Cities Confront the Nazi Past, edited by Gavriel D. Rosenfeld and Paul B. Jaskot
The Politics of Sociability: Freemasonry and German Civil Society, 1840–1918, Stefan-Ludwig Hoffmann
Work and Play: The Production and Consumption of Toys in Germany, 1870–1914, David D. Hamlin
The Cosmopolitan Screen: German Cinema and the Global Imaginary, 1945 to the Present, edited by Stephan K. Schindler and Lutz Koepnick
Germans on Drugs: The Complications of Modernization in Hamburg, Robert P. Stephens
Gender in Transition: Discourse and Practice in German-Speaking Europe, 1750–1830, edited by Ulrike Gleixner and Marion W. Gray
Growing Up Female in Nazi Germany, Dagmar Reese
Justice Imperiled: The Anti-Nazi Lawyer Max Hirschberg in Weimar Germany, Douglas G. Morris
The Heimat Abroad: The Boundaries of Germanness, edited by Krista O'Donnell, Renate Bridenthal, and Nancy Reagin
Modern German Art for Thirties Paris, Prague, and London: Resistance and Acquiescence in a Democratic Public Sphere, Keith Holz
The War against Catholicism: Liberalism and the Anti-Catholic Imagination in Nineteenth-Century Germany, Michael B. Gross
German Pop Culture: How "American" Is It? edited by Agnes C. Mueller
Character Is Destiny: The Autobiography of Alice Salomon, edited by Andrew Lees
Other Germans: Black Germans and the Politics of Race, Gender, and Memory in the Third Reich, Tina M. Campt
State of Virginity: Gender, Religion, and Politics in an Early Modern Catholic State, Ulrike Strasser
Worldly Provincialism: German Anthropology in the Age of Empire, edited by H. Glenn Penny and Matti Bunzl
Ethnic Drag: Performing Race, Nation, Sexuality in West Germany, Katrin Sieg

For a complete list of titles, please see www.press.umich.edu

Work, Race, and the Emergence of Radical Right Corporatism in Imperial Germany

Dennis Sweeney

THE UNIVERSITY OF MICHIGAN PRESS

Ann Arbor

Published in the United States of America by
The University of Michigan Press
Manufactured in the United States of America
∞ Printed on acid-free paper

2012 2011 2010 2009 4 3 2 1

A CIP catalog record for this book is available from the British Library.

Library of Congress Cataloging-in-Publication Data

Sweeney, Dennis, 1961–
Work, race, and the emergence of radical right corporatism in imperial Germany / Dennis Sweeney.
p. cm. — (Social history, popular culture, and politics in Germany)
Includes bibliographical references and index.
ISBN-13: 978-0-472-11678-2 (cloth : alk. paper)
ISBN-10: 0-472-11678-9 (cloth)
1. Corporate state—Germany—History. 2. Steel industry and trade—Germany—Saarland. 3. Iron and steel workers—Germany—Saarland. 4. Industrial relations—Germany—History. 5. Employee rights—Germany—History. 6. Paternalism—Germany—History. 7. Authoritarianism—Germany—History. 8. Conservatism—Germany—History. 9. Germany—Politics and government—19th century. I. Title.
HD3616.G42S94 2009
338.6—dc22 2008033530

For my parents, Mary and Don Sweeney

Contents

Abbreviations

ARV	Allgemeine Arbeiter-Rechtsschutzverein
AVBS	Arbeitgeberverband für das Baugewerbe und die verwandten Berufe der Saargegend
AVSI	Arbeitgeberverband der Saarindustrie
BAT	Bistumsarchiv Trier
BdI	Bund der Industriellen
BF	*Der Bergmannsfreund*
BM	*Bürgermeister*
CMV	Christlicher Metallarbeiterverband Deutschlands
CVDI	Centralverband Deutscher Industrieller
DAF	Deutsche Arbeitsfront
Dinta	Deutsches Institut für technische Arbeitschulung
DSVP	Deutsch-Saarländische Volkspartei
EW	*Evangelisches Wochenblatt*
GA	*General-Anzeiger*
GN	*Gewerkschaftliche Nachrichten*
HDA	Hauptstelle Deutscher Arbeitgeberverbände
LAS	Landesarchiv Saarbrücken
LHAK	Landeshauptarchiv Koblenz
LR	*Landrat*
MBZ	*Malstatt-Burbacher Zeitung*
NEA	Neunkircher Eisenwerksarchiv
NSDAP	Nationalsozialistische Deutsche Arbeiterpartei
NSGB	*Neue Saarbrücker Gewerbeblatt*
NSZ	*Neue Saarbrücker Zeitung*
NVZ	*Neunkircher Volkszeitung*
OP	Oberpräsident
RDI	Reichsverband der deutschen Industrie
RP	Regierungspräsident

SAD	Stadtarchiv Dudweiler
SAS	Stadtarchiv Saarbrücken
SAN	Stadtarchiv Neunkirchen
SGB	*Saarbrücker Gewerbeblatt*
SGVDESI	Südwestliche Gruppe des Vereins Deutscher Eisen- und Stahlindustrieller
SIH	*Saarindustrie und Handel*
SJSVZ	*St. Johanner-Saarbrücker Volkszeitung*
SJVZ	*St. Johanner Volkszeitung*
SJZ	*St. Johanner Zeitung*
SP	*Saarpost*
SPD	Sozialdemokratische Partei Deutschlands
SBMG	Stadtbibliothek Mönchengladbach
SW	*Saarwacht*
SWDWZ	*Südwestdeutsche Wirtschaftszeitung*
SZ	*Saarbrücker Zeitung*
VDA	Verein Deutscher Arbeitgeberverbände
VDESI	Verein deutscher Eisen- und Stahlindustrieller
VW	*Volkswacht*
VWGWISI	Verein zur Wahrung der gemeinsamen wirtschaftlichen Interessen der Saarindustrie
ZfS	*Zeitschrift für Sozialwissenschaft*

Acknowledgments

I began this book some years ago in the context of my dissertation research at the University of Michigan, where I learned from the very best and acquired numerous intellectual debts. Above all, I thank my doctoral advisor, Geoff Eley, whose knowledge of German and European history, generosity, patience, sense of humor, and unfailing support continue to amaze me; Kathleen Canning, who commented extensively on my work and has influenced and supported it in so many ways; and Bill Sewell, whose quiet pedagogy, theoretical clarity, and work in French labor and social-cultural history were key sources of inspiration. My other committee members and teachers at Michigan—Peggy Somers, Ron Suny, Nick Dirks, Michael Schneider, and Heidrun Homburg—also deserve my most heartfelt thanks. I am grateful to several former colleagues at the College of New Jersey: Tom Allsen, Dan Crofts, the late Alan Dawley, Joanne Gross, and John Karras, who threw a going-away party that I will never forget. At the University of Alberta, I would not have survived the cold winters without my friend and colleague John Kitchen.

I have also benefited from conversations with a wide range of friends and colleagues, many of whom have commented on my work and helped me clarify my thinking on all kinds of issues that appear in this book. They include David Crew, Young Sun Hong, Rainer Horn, Larry Frohman, Kristin McGuire, Jean Quataert, Markus Reisenleitner, Steve Soper, George Steinmetz, and Peter Hohendahl. I am especially grateful to Keith Nield and Janet Blackman, editors of *Social History,* who have been supportive of my work over the years. I also thank Chris Hebert, Marcia LaBrenz, and the two anonymous readers for the University of Michigan Press for their insightful and constructive comments on this book. Larry Cheever made the superb map.

In Germany and North America, I benefited from the assistance of numerous archivists, historians, and librarians at the Landesarchiv Saarbrücken, Stadtarchiv Saarbrücken, Stadtsbibliothek Saarbrücken, Stadtbibliothek Mönchengladbach, Bistumsarchiv Trier, Stadtarchiv Neunkirchen,

Neunkircher Eisenwerksarchiv, Landeshauptarchiv Koblenz, Bundesarchiv in Koblenz, and University of Alberta. I especially thank Michael Sander of the Landesarchiv Saarbrücken, who shared with me his own research on the confessional labor movement in the Saar; Stefan Leiner, who gave me copies of his own work on urbanization and migration in the region; Heinz Gillenberg, who assisted me at the archives in Neunkirchen; and Klaus Ries, a historian of the Saar region with whom I shared many conversations. I thank Celia DeCastro at the University of Alberta for guiding and executing my interlibrary loan quests with such patience and thoroughness.

The research for this book was supported by several organizations, and parts of it have appeared in print before. The German Academic Exchange Service (DAAD) funded my first trip to archives in Germany. Subsequent trips were funded by the Rackham Graduate School at the University of Michigan. I am grateful to the DAAD and the National Endowment for the Humanities for a summer grant that supported my participation in a seminar entitled "Theories of the Public Sphere" at Cornell University, where I found time to think more about the significance that the debates over *Öffentlichkeit* held for my own work. A sabbatical from the University of Alberta allowed me to finish the manuscript. I gratefully acknowledge the Taylor and Francis Group for permission to reprint, in chapter 5, sections of my article "Work, Race, and the Transformation of Industrial Culture in Imperial Germany," *Social History* 23 (1998): 31–62. I thank Cambridge University Press for permission to reprint—in the introduction, chapter 6, and the epilogue—sections of my article "Corporatist Discourse and Heavy Industry in Wilhelmine Germany: Factory Culture and Employer Politics in the Saar," *Comparative Studies in Society and History* 43 (2001): 701–34. Chapters 2 and 4 include sections of my article "Liberalism, the Worker, and the Limits of Bourgeois *Öffentlichkeit* in Wilhelmine Germany," *German History* 22 (2004): 36–75. I am grateful to the German History Society for permission to reprint those sections.

I also thank friends and family for their support over these past many years, especially Steve Soper, for all of our discussions about German and Italian history and life in general, and my colleague and comrade Geoff Eley, whom I could never thank enough. I am very grateful to my parents, Mary and Don Sweeney, for all of their unconditional encouragement and understanding over the years during which I wrote this book. Finally, I thank my partner, Deb Ferkes, who directly assisted in the preparation of this book in numerous ways and who inspires me everyday with her intelligence, honesty, enthusiasm, and love.

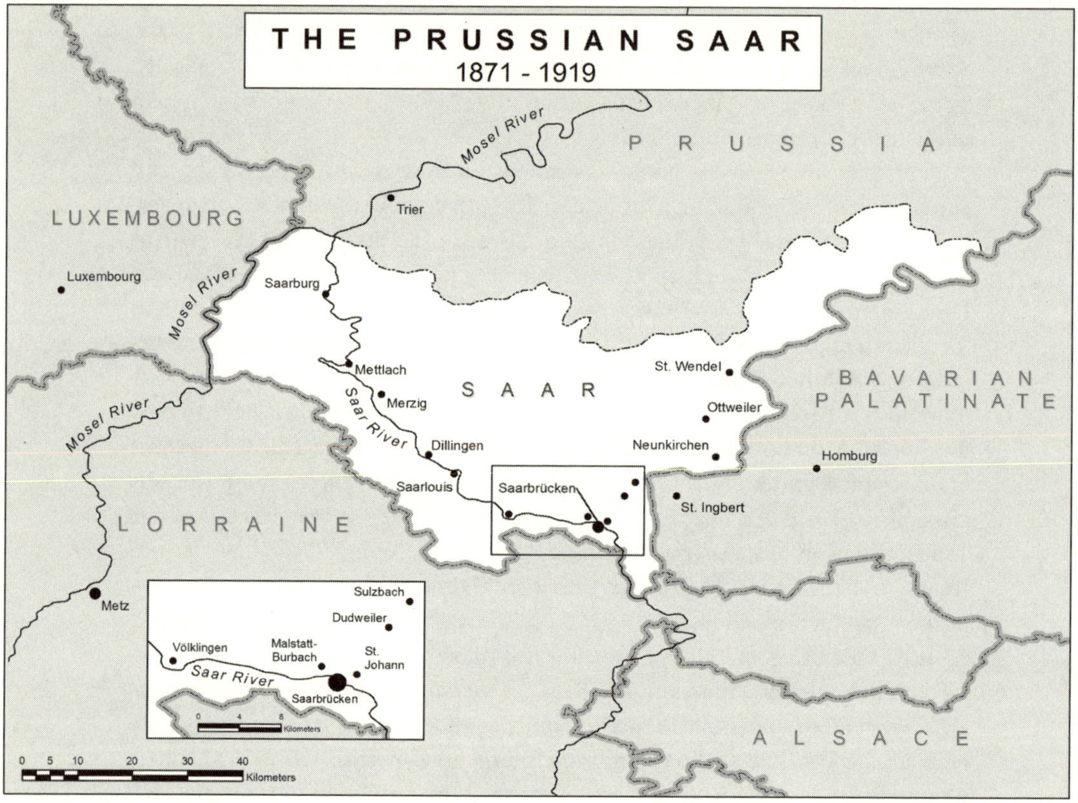

THE PRUSSIAN SAAR
1871 - 1919
PRUSSIA
Mosel River
Trier
LUXEMBOURG
Luxembourg
Saarburg
Mosel River
Mettlach
SAAR
St. Wendel
BAVARIAN
PALATINATE
Merzig
Saar River
Ottweiler
Mosel River
Dillingen
Neunkirchen
Homburg
Saarlouis
Saarbrücken
St. Ingbert
LORRAINE
Metz
Sulzbach
Dudweiler
Völklingen
Malstatt-
Burbach
St.
Johann
Saar River
Saarbrücken
Kilometers
ALSACE
0 5 10 20 30 40
Kilometers

Introduction

In late September 1905, a number of leading social reformers and representatives of German heavy industry met in Mannheim at the annual general meeting of the Association for Social Policy (Verein für Sozialpolitik), the preeminent social reform organization in imperial Germany, for a debate over "labor relations in private large-scale industrial enterprises."[1] Against the backdrop of rising labor militancy—especially the recent miners' strike in the Ruhr—and the increasingly aggressive antiunionism of German industrialists, most of the reformers at Mannheim called for fundamental institutional changes to the industrial workplace, changes that would allow for negotiations between employers and employees over the conditions of work and curb the excessive power of German industrialists over their workers. The sociologist Max Weber, responding to the comments of heavy industry spokesman Alexander Tille, focused his criticism on labor relations in the Saar Valley, a region of coal mining and iron and steel manufacturing in the southwest corner of the Prussian Rhine Province. The Saar was best known as the home of the late iron and steel magnate and Free Conservative politician Carl Ferdinand von Stumm-Halberg, once imperial Germany's most vocal exponent of the paternalist factory regime, which combined far-reaching disciplinary and moralizing work rules with extensive company welfare benefits, from wage bonuses to housing. The work rules in Saar factories, Weber maintained in Mannheim, resembled "police jargon." As such, he argued, they reflected "German traditions": the striving for the "appearance of power" rather than real power; the "petty bourgeois thrill of being a gentleman" (*spiessbürgerliche Her-*

renkitzel) revealed in the actions of the industrial paterfamilias, who had to show "those under him" that he "has something to say"; and the authoritarianism inherent in the paternalist "system" generally.

> I myself . . . know very well the Saar region and the suffocating atmosphere which this system has created there—not for you, Dr. Tille, but for others, and not just for workers but for any who dare to be politically active in any way that displeases these men [i.e., Saar employers]. Gymnasium teachers, officials, and all with whom I came in contact at the time confirmed that anyone considered a civil servant—up to and including the *Oberpräsident* [Provincial Governor]—dances to the tune of these men, that any independent point of view risked the threat of transfer or reprimand.[2]

In ways that summarized the views of most liberal reformers, Weber argued that the paternalist factory regime was founded on a "contract of subordination" that violated "modern notions of legality." Much like the "authoritarian" Prussian state, it turned workers into subjects or "rabble" (*Kanaille*) rather than "honorable citizens." In this way, Weber interpreted the labor policies and factory regimes in Saar heavy industry as relics from a premodern or feudal past.

Nevertheless, the defense of employers mounted in Mannheim by industry spokesmen Tille and Eugen Leidig, deputy chairman of the Central Association of German Industrialists (Centralverband Deutscher Industrieller, hereafter CVDI), contained little of the moralizing claims of employers like Stumm or the familiar features of Saar paternalism—its description of the employer as economic and moral "provider" for workers, its repeated invocation of a "personal relationship between employer and employee," and its support for state *Sozialpolitik.* Instead, Tille assailed what he regarded as the unscientific and "social-moral" claims of the reformers, their calls for more state welfare programs and institutional mechanisms for workplace bargaining, and he defended employer prerogative in terms of a scientific and technocratic vision of work relations and economic efficiency. The terms and conditions of work in large industrial concerns, Tille maintained, could not be set according to the inapplicable, artificial, egalitarian, homogenizing, and ethical considerations that motivated state-sponsored welfare and workplace regulation. Rather, they should be determined by natural biological and economic laws, which pitted workers of differing bioracial capacities against each other in healthy competition, and by the quality of a worker's "performance" (*Leistung*) on the job. In a technocratic formulation that stressed the importance of the "entrepreneur" (*Unternehmer*) as a general category of producer, Tille also criticized state-sanctioned arbitration bodies and increasing state regulation of the industrial workplace as threats to "economic freedom"—"our greatest good"—and to the "personal freedom" of employers and workers equally to choose with whom and under what conditions they would work. This departure from the paternalist language of Stumm would be incorporated three years later into a more systematic corporatist discourse about a racially defined social order composed of "occupational estates" (*Berufsstände*), in which a "social aristocracy" of "productive employers" was called on to defend its prerogatives and the conditions of capitalist profitability against trade unions, irresponsible social reformers, meddlesome political parties, and an interventionist welfare state.

Leidig's response to the reformers at Mannheim also departed from the once standard paternalist references and claims of industrialists and emphasized the changed context of labor relations in German heavy industry after 1900. The CVDI deputy chairman defended the freedom of the "purely economic" labor contract and the prerogatives of employers in heavy industry in terms of the organizational and economic imperatives of industrial production in an age of joint stock companies and global markets. He argued that "inside the factory the authoritarian position of the employer should be protected and maintained not . . . in the sense of a tsarist autocracy but in the sense of an organization," which, like any organization involved in a "difficult competition," must be led by a single, "unitary will." In Leidig's view, the employer (*Unternehmer*) was caught in a global competition for markets, bound to a much wider community of capital investors, and thus more restricted than ever in his room for maneuver than paternalist employers who once owned their own factories. Moreover, he found himself increasingly on the defensive—and not at all the "stronger" partner in the wage relationship—in the "new epoch of struggle" between workers and management defined by mass unionism and the regional and national organization of employers after 1900. Leidig identified the encroachments of trade unions, social reformers, and state regulators as looming threats, architects of an impending "socialist state," but he concentrated on two weapons available to workers as the principal threats to employers.

> The one is the universal and equal franchise, which has been put in the hands of workers; the other is public opinion and the press, which is likewise today used in a comprehensive way by the workers. With their silent, steady, pressing authority, these institutions, which redound to the benefit of workers in no other nation in the civilized world to this extent, force all political parties to engage in the struggle for the votes of workers. Their enduring effect is to place German employers under an influence that, each and every time they impose a disciplinary sanction, forces them to consider most carefully the question: is it absolutely necessary to reject the demands of the worker?[3]

Like Tille's, Leidig's arguments about the organization of labor conflict, mass politics, and the growing power of public opinion registered the corporatist restructuring of German political economy—understood here in terms of the increasing interpenetration of economic and party-political spheres—and pointed to a range of new concerns that transformed employer discourses about work and industrial organization in late imperial Germany.

German historians have not yet fully appreciated the changing terms of this debate or political struggle over factory organization and the transformation of employer discourses in Germany during the decade before 1914. Until recently, they had been mainly preoccupied, like Max Weber, with the allegedly "premodern" or "feudal" origins of paternalist or corporatist labor policies in German heavy industry—an interpretation that derived its staying power from the *Sonderweg* (special path) thesis about the long-term origins of Nazism. Consolidated during the 1960s and 1970s by the practitioners of "historical social science" (*Gesellschaftsgeschichte*), a largely Weberian reformulation of modernization theory applied to social historical writing, the *Sonderweg* thesis stressed the fundamental divergence of German sociopolitical development from the developmental trajectories of the "West." It emphasized the fatal disjuncture between a highly advanced industrial economy and a backward, illiberal political system and culture, which prevented the "normal" evolution of parliamentary democracy out of the social energies released by industrial capitalism.[4] The "preindustrial" elites and social groups that prevented this model of development came from the aristocracy, the civil bureaucracy, and the army, but they were joined by industrialists like Stumm and Krupp, as well as a younger "cohort" of Ruhr "manager-entrepreneurs" like Emil Kirdorf, Fritz Baare, or Hanns Jencke, who aped the illiberal values of traditional elites in political outlook and lifestyle.[5] Their desire to remain "master of the house" (*Herr im Hause*) in the factory, according to Han-Ulrich Wehler, reflected a "residual style of traditional leadership" that rejected capitalist "market rationality" and the economic benefits that came with "modern" managerial practices and the acceptance of trade unions.[6] This traditional attitude, so this argument runs, was also expressed in the new managerial rationalities deployed in the technologically advanced industries of the "second industrial revolution," which drew on "preindustrial" bureaucratic practices and feudal hierarchies,[7] and in employer yearnings for a corporatist order of occupational "estates," which derived from a specifically German tradition of romantic "antimodernism" that rejected modern "liberal, secular, and industrial civilization" or "conservative feudal value orientations" of small retailers, artisans, and industrialists from domestic heavy industry who sought protection from the competitive pressures of a modern capitalist system.[8] From this perspective, paternalist and "corporatist-authoritarian" coal and steel barons stifled the budding prospects of "capitalist democracy" in the Wilhelmine era; contributed to the "willful, planned undermining of the sociopolitical compromises . . . of the Weimar Republic"; and helped to pave the way for the antimodernist, authoritarian appeal of National Socialism during the late 1920s and early 1930s.[9]

During the 1980s, German historians decisively challenged this perspective on the preindustrial attitudes and behaviors of employers and have since emphasized the "modernity" of industrial relations and the economic rationality of paternalist and corporatist labor policies from the late nineteenth and twentieth centuries.[10] Geoff Eley and David Crew were the first to debunk systematically the implicit teleologies of the *Sonderweg* thesis in relation to employers like Stumm and Krupp, especially the a priori normative assumption that employers in modern capitalism would necessarily come to accept the benefits of trade unions and social partnership. They argue that industrial paternalism is best understood as a common and instrumental "way of controlling labor costs by preventing workers from organizing and striking"; maintaining a compliant labor force by providing social benefits and housing, particularly in regions lacking urban amenities and infrastructure; and preempting state welfare and regulatory intervention into the industrial workplace and their associated restrictions on employer prerogative.[11] In this context, according to Eley, the paternalist organization of factory production accorded with a historically "specific form of capitalist rationality" that was "immanent to the monopoly phase of German capitalism"; it was thus a product not of a "bygone feudal era" but of a "set of relations generated by capitalism itself" during the Wilhelmine era.[12] Subsequently, historians have stressed the "rational market calculation" that inspired paternalist company welfare schemes and antiunion measures and the basic continuity of paternalist policies in heavy industry from the nineteenth century to the Nazi era.[13] Mary Nolan's major study of industrial rationalization during the Weimar era has similarly emphasized the economic function of right-wing corporatist visions of "factory community" and efforts to "engineer" an obedient, antiunion "new worker." If these initiatives were "archaic" or "not the most modern," Nolan argues, they nevertheless served the economic interests of employers by promoting "productivity and profitability."[14]

This study critically interrogates this emphasis on the continuities of work relations and production regimes in German heavy industry from the Kaiserreich to the Third Reich, by means of a case study of discourses about work and social order in the industrial Saar, the focus of Max Weber's critique and a common reference point for much subsequent social-historical interpretation. In an attempt to make sense of the new vocabulary of Tille and Leidig and of the wider assumptions that informed their statements at the Mannheim conference, it identifies an important discontinuity in industrial discourses or ideologies of workplace and social organization during the Wilhelmine era: namely, the shift from a paternalist discourse of work and social relations, structured in a moralizing and gendered metaphor of a factory "family" and anchored in rigid work rules and extensive company social programs, to a corporatist discourse of industrial social organization, which linked a bioracial schema of technocratic management to a wider vision of sociopolitical order based on representation by occupational groups or "productive estates" (*Berufsstände*). As this study demonstrates, this corporatist discourse framed a new and forward-looking authoritarian understanding of industrial-capitalist society, an understanding that converged with core elements in the ideological configurations of the radical Right in late Wilhelmine Germany and anticipated subsequent discourses about industrial work and social organization under National Socialism.

In view of its broader scope and relevance, therefore, this study of Saar factory regimes and political cultures enters into recent historiographical debates over the onset of "modernity" and the generative contexts of right-wing ideologies of work and social order in Wilhelmine Germany. The social science historians understood employer discourse and labor relations in heavy industry in terms of the continuity of "preindustrial" traditions, and revisionist critics interpret them in relation to the continuity of capitalist imperatives and rationalities. This study offers a cultural-historical reading of multiple and competing industrial discourses—including paternalism, social Catholicism, Protestant-liberal reform, social democracy, and corporatism—and their historically specific productivity, that is, their varying capacity to shape work relations and production regimes in Saar heavy industry. In this way, it offers two challenges to existing historiographical interpretations of employer ideology in German heavy industry. First, in response to arguments about either the "feudal" nature or strictly economic logic of paternalist labor relations, which have focused on figures like Stumm and Krupp in particular, this study seeks to demonstrate the bourgeois character of company paternalism as both a rational way to conceptualize and organize capitalist production and a moralizing discourse about workers' comportment and familial order. Second, its focus on the rise of corporatist schemes of industrial and social order not only emphasizes an important transformation of heavy industrial labor policy and workplace organization during the late Wilhelmine era; it also calls attention to the generative ideological context and emergence of right-wing constructions of the economic "modern" in German heavy industry: namely, the increasing penetration of technocratic assumptions, scientific design, and racist intentions within managerial thought and practice; the greater reliance on impersonal norms and internalized disciplinary technologies, rather than overt managerial injunctions, in the regulation of factory labor; and the turn to wider schemes of corporatist social reorganization in response to organized conflict over the shape of the industrial workplace, more comprehensive forms of state labor regulation and social policy, and global economic competition.[15]

In this way, this study identifies a set of historical articulations revealed in the prewar activities of industrialists from Saar and other German locations, with important consequences for German history. It reveals connections between capitalist managerial practices, technocratic models of workplace organization, comprehensive forms of collective industrial interest representation, and antidemocratic and racist corporatist schemes of sociopolitical order that were forged in the ideological common ground between important sectors of German heavy industry and the radical Right on the eve of war in 1914. These schemes were disseminated more broadly during the Weimar era and ultimately taken up and rearticulated by the National Socialists during the 1930s. In exploring these broader consequences, therefore, this study attempts to reinvigorate the rather dormant efforts to explore the historical genealogies of Nazism in relation to industrial conflict, employer politics, and the development of industrial capitalism in Germany after 1890.

Debates over the connections between heavy industry and the German Right and between capitalism and Nazism were vibrant from the late 1960s to the 1980s, but interest in historical genealogies of this kind has been muted by two historiographical trends during the last two decades. The first is the general reluctance of social and cultural historians since the 1980s to formulate explicit arguments about the causal connections between the politics and labor policies of German heavy industry and the political Right during the late Wilhelmine period, the Weimar Republic, and the Third Reich, in the wake of the successful critique of the *Sonderweg* thesis. For the most part, they have not gone beyond emphasizing the role of industrialists in undermining the legitimacy of the Weimar Republic, identifying the similarities between Weimar-era managerial strategies focused on the "factory community" (*Werksgemeinschaft*) and Nazi labor policies during the 1930s, or redefining fascist industrial policy as part of a generalized cultural "crisis of modernity."[16]

The second historiographical trend has involved the gradual acceptance of the "primacy of politics" thesis, which has encouraged interpretations of the collapse of Weimar, the rise of National Socialism, and histories of the Third Reich that no longer assign causal significance to labor and class relations, employer politics, and capitalism. The most influential formulation of this thesis came from Tim Mason, who argued as early as 1966 that Nazi leaders "broke the economic and political supremacy of heavy industry" by means of political decisions about rearmament and state intervention into the economy; became "independent of the influence of the economic ruling classes" and even devised policies that "ran contrary to" the interests of those classes; and prevented industrialists and their representatives from having any "decisive" influence on Nazi foreign diplomacy, war aims, and efforts to create a "new imperial order in Europe."[17] Studies of individual German firms and "businessmen" during the Weimar and Nazi eras have since portrayed the majority of industrialists as opponents of Nazism, as ineffectual political actors who were powerless to influence events leading to the collapse of the Weimar Republic and the rise of the Nazis, or as unethical businessmen opportunistically and self-interestedly, rather than ideologically, complicit in the social and economic policies of the Third Reich.[18] General analyses of the dirigiste "Nazi economy" also tend to downplay the significance of the private sector and capitalist developments to the fate of the Nazi regime, and some even emphasize what they regard as the fundamental contractions between the "Nazi economy" and capitalism.[19] Only very recently have historians once again begun to acknowledge the apparent "congruity of interests between state and economy," the "room for independent activity" (*Betätigungsspielräume*) left to German industrialists during the Third Reich, and the disturbing conjunction of racial and economic imperatives in the Nazi genocide.[20]

This study will respond to these two larger historiographical trends as it attempts to rethink the historical connections between industrial class relations, employer discourses, and right-wing politics in the genealogies of Nazism. It will do so by drawing on the theoretical frameworks of cultural theory and the insights of the "cultural turn" in labor history in order to explore the changing ideological discourses about work and sociopolitical order in the prewar Saar region. In particular, this study will draw attention to the conditions of emergence of the racially charged corporatist project in Saar and German heavy industry and its subsequent connections to the racist labor policies and wider corporatist social imaginary of the National Socialists in the 1930s.

The Cultural Turn and Ideological Discourse

The linguistic or cultural turn has emerged from a wide variety of often divergent theoretical developments across the human sciences during the last two decades, including Gramscian Marxism, women's history and feminist theory, cultural anthropology, cultural studies, historical sociology, literary deconstruction, and poststructuralist philosophy. In a highly overdetermined way, it produced a critique of the "materialist model of social determination," which treats political languages as the more or less entailed outcomes of autonomous material-economic processes, and turned instead to analysis of the constitutive role of the discourses, languages or representations of labor, the meanings of work, market cultures, and political factors in the very formation of economic order, factory organization, workplace struggles, labor movements, and processes of class formation.[21] In German historiography, the cultural turn arrived belatedly, via feminist history and "the history of everyday life" (*Alltags-*

geschichte).[22] But during the last ten years, it has inspired numerous studies of working-class formation, the political languages of workers, regionally varied production regimes or "industrial orders," the "micropractices" of factory production, and the factory as a "field of action" (*Handlungsfeld*). These studies emphasize the cultural or discursive dimensions of economic practices and social relations during the nineteenth and twentieth centuries.[23]

Nevertheless, this emphasis on the constitutive role of discourse has been largely absent in debates among German historians over the meanings and import of industrial paternalism and corporatism and the relations between ideology and the economy from the Kaiserreich to the Third Reich. Despite their fundamental and productive differences, social science historians and their critics who study industrial relations and employer politics during the late nineteenth and early twentieth centuries share certain analytical claims and ontological assumptions about the binary relationship between language or discourse and material reality: they tend to treat languages as relatively fixed systems of meaning that (should) reflect underlying or extralinguistic economic conditions or social structures. In this way, they interpret the meaning of employer discourse in terms of the extent to which it "rationally" corresponds with the general trajectory of capitalist industrialization, understood as an objective material process occurring outside the realm of representational practice or language. This kind of approach neglects the historical productivity or effectivity of discourse and imputes an intrinsic logic to economic and social development, which calls forth cultural or ideological responses; it thereby measures the actions and languages of historical actors in specific "social settings" against an "endogenous directionality or *a priori* definition of rational action" and thought.[24]

By contrast, recognizing the different and competing languages of industrial social order in the Saar, this study draws on the insights produced in discussions about discourse and ideology within Gramscian cultural studies, which treat discourse as a social activity and thus recognize its historical productivity and variability.[25] In the wake of Althusser's theory of ideology and Bakhtin's reflections on "dialogism," practitioners of cultural studies proposed a definition of ideological discourse as a historically evolved and socially organized ensemble of material signifying practices—a "connotative field of reference" that interpellates subjects and defines their relation to social order.[26] This definition offers a number of advantages for this study of labor relations and industrial politics. First, by coupling the term *ideology* to the term *discourse,* this definition draws attention to the differential workings of power in language and the hierarchical relations between discourses in different historical and social settings.[27] Second, it points to the relational or "dialogical" dimensions of discourses, their necessary orientation toward and implication in other discourses. In these ways, this definition offers a useful way of analyzing the historical dynamics of discursive articulation—understood here in terms of both expression in language and connection (or linkage) between linguistic elements—as the ongoing and contingent process of hegemonic struggle and repositioning that takes place within and between discourses as they seek to define subjects and order the social world.[28]

Third, this approach permits analysis of the ways in which ideological discourses articulate to other material practices and institutions in response to both previous social-historical and recent cultural-historical paradigms. If social historians and many cultural historians still reify conceptual distinctions between the "material" and the "discursive" and thereby fail to interrogate the constitutive role of discourse in the very creation of the "economic structures" of "capitalist development," other cultural historians attempt to subsume all material practices and institutions into their definition of discourse without distinguishing between what are more usefully understood analytically as different dimensions of social life.[29] They do so by insisting, with Joan Scott, that there is no "opposition between material life and political thought" and by then proceeding to examine published texts as instantiations of discourse.[30] But this kind of approach cannot fully explain two aspects of ideological discourse crucial for the historian of industrial class relations in Germany in the late nineteenth and twentieth centuries: namely, why some discourses about industrial work and social order became dominant while others remained subordinate or even disappeared in particular historical eras or conjunctures, as well as how such discourses came into being and were transformed over time.[31] To understand these aspects of paternalism and corporatism in the Saar, this study relies on an analytical distinction between discourses and other material practices and recognizes the imperative to examine their necessary interarticulations in any lived social context. It proposes that the signifying power of an ideological discourse is best understood in terms of both its connotative resonances and its materialization in other social practices and institutions.

This study therefore questions assumptions about the semantic fixity and logical coherence of ideological discourses over time and fixed ontological distinctions between the economic and the ideological. Following Stuart Hall, it argues that our interpretations of ideological discourses about work and social order, particularly the meanings of Saar paternalism and corporatism, should avoid a priori assumptions about their necessary "'logics' of arrangement" and approach them instead as historically contingent formations whose connotative references and principles of articulation were variable. From this perspective, rather than identifying stable discourses and their long-term semantic unities, it makes more sense to explore how different linguistic "elements come, under certain conditions, to cohere together within a discourse" and "how they do or do not become articulated, at specific conjunctures, to certain political subjects" and institutional forms.[32] To suggest that discourses, as modes of social activity, are made in the course of social struggles and therefore undergo processes of "modification and development" over time is not to claim that they are ephemeral or permanently in flux.[33] Rather, this study argues that precisely the extent to which certain discourses became articulated to other social practices secured their connotative power and allowed them to become structurally and "materially effective" in the Saar in ways that other discourses were not. It thereby proposes a model of cultural historical practice that attempts to analyze social and political change in the Saar by tracing dynamic and mutually determining relations between discourses and other practices and domains of social life over time: that is, the historical processes by which a dominant paternalist discourse about economic and social order was forged in practices of speech and writing; materialized in social relations, public institutions, and built environments in ways that set the main parameters for political-ideological struggle in the region; subsequently challenged by alternative discourses about work and social order; and ultimately supplemented and replaced by a new corporatist discourse with its own semantic articulations and range of material connections and effectivities during the prewar decade.

The Public Sphere and the State

A wider focus of this kind requires paying attention to what Bakhtin referred to as the "social life" of discourses—in this case, their articulations and deployments across a range of sites, which go far beyond the industrial workplace and into the translocal "public sphere" and the agencies of the imperial German state.[34] This study explores the formation of ideological discourses about work in a wide range of texts and institutions—including factory work rules; management circulars; workplace rituals; company brochures; and the statutes, speeches, meeting minutes, and memoranda produced by voluntary associations, city and town councils, social reform organizations, and trade unions—as well as the everyday activities, organizational structures, social relations, and modes of political deliberation that characterized the industrial workplace, social clubs and voluntary associations, local governmental bodies, and political organizations across the Saar region. In addition, it examines the role of the state—as a locus and framework of workplace regulation and social welfare and a repository of legal-constitutional guarantees of political rights—in the organization of work and the modalities of politics in the Saar. It thus explores the transformation of factory regimes and relations of class hegemony in the Saar in the complex interplay between the arenas of the industrial workplace, the public sphere, and the state. This perspective emerges from two sets of debates related to Jürgen Habermas's theory of the public sphere and the role and function of the German state.

Historians have increasingly been drawn to Habermas's theory of the bourgeois public sphere in order to explore vital and once neglected forms and arenas of political activity and association, outside of the domains of the state and economy, from the eighteenth to the twentieth centuries. In the process, they have critically but sympathetically challenged Habermas's idealized theorization of the bourgeois public sphere—especially its normative investment in a domain, characterized by universal access, of self-correcting rational-critical discourse beyond or outside of politics—and proposed its redefinition as, in Geoff Eley's words, "the structured setting where cultural and ideological contest or negotiation among a variety of publics takes place."[35] This important reworking of the public sphere as a plural domain, or set of multiple and interconnecting public spheres in which various social groups articulate identities and stake political claims, registers both the historically specific exclusivity of the bourgeois public sphere and its democratic possibilities in the nineteenth and twentieth centuries.[36] This study will similarly examine the formations of *Öffentlichkeit* across the cities and small towns of the Saar as the crucial arenas in which the organization of work and the balance of social forces in the Saar were secured but also contested by Saar workers. It thus enters into a sustained critical engagement with production-centered studies in labor history by relocating the struggle over labor relations and social order to the arenas of the public sphere.

Yet these revisions to Habermas's theory tend to treat the public sphere as a self-contained and even social space (to which historical actors attempt to gain access) and thus to obscure the extent to which *Öffentlichkeit* was structured in relations of social hierarchy and economic inequality in the late nineteenth and early twentieth centuries. Even those studies attuned to the uphill struggle of subordinate groups seeking access to "public" debate have not fully appreciated the inequalities and limitations intrinsic to the bourgeois public sphere, its constitutive (not incidental) relation to the "material sphere of everyday life, the social conditions of production" and reproduction.[37] By contrast, this study pays attention to the material and social-structural parameters of the public sphere: the matter of who owned or controlled the principal media of public communication and who was allowed to participate in its deliberations, as well as the processes by which hegemonic definitions of *Öffentlichkeit* became embodied in certain kinds of institutions, deliberative routines, and discursive norms. This emphasis is vital for any analysis of public life in a region where the owners and directors of heavy industrial concerns, figures like Stumm in particular, deployed their considerable economic and financial power to curb or even silence the expression of oppositional or alternative speech and activity.

Moreover, these reworkings of Habermas's theory generally adhere to a social or "spatial" model of the public sphere, as a domain of institutions and organizations, which tends to overlook a crucial second definition of *Öffentlichkeit:* the collective subject or "public" forged in political debate. As Harold Mah perceptively argues, they have neglected analysis of the process by which the multiple voices of political life are transformed into a "phantasmatic" unity, the fiction of a unified public. This kind of analysis would refocus our attention on the central discursive strategy of legitimation intrinsic to the bourgeois public sphere: namely, the claim to speak in the name of "the public." When political actors invoke the public as collective subject in this way, they presume not the copresence of multiple subjects and rationalities in formation but the fantasy of an already existing, unitary mass subject in possession of universal reason.[38] Accordingly, this study examines the public sphere in the Saar not as a neutral, independent domain of public reason but as a highly contested ideological construction, not only forged in struggles over the public as a social domain, the conflict between employers and workers over the formation of new associations, the renting of meeting halls, and the right to speak publicly, but also wrought in debates over who counted as "the public" and what the public could be said to authorize. It examines thereby the actual references to and the struggle over the meaning of the term *Öffentlichkeit* in the Saar during the Wilhelmine period. In this context, the public sphere emerged as a complex and often contradictory "cross-citational field" of discursive interaction, a general "criterion" of openness, and a collective political subject, toward which employers, social reformers, state officials, and many workers oriented themselves and through which the terms and conditions of industrial work and the shape of labor relations were contested in the Saar after 1890.[39]

Finally, by contrast with Habermas's influential treatment of the bourgeois public sphere as a discrete domain of discursive interaction, this study explores the complex and generative interconnections between the industrial workplace, the public sphere, and the imperial German state.[40] This requires drawing on recent developments in state theory in relation to German historiography on the Kaiserreich. German historians, like Habermas, have long treated the state as a discrete, unitary, autonomous, and overwhelmingly repressive entity, composed of an identifiable set of institutions and actors, whose concerns about governance and "policing" extend back to the late medieval and early modern periods and constituted the premodern core of the antidemocratic Wilhelmine state. Drawing on the Marxist theory of Nicos Poulantzas, Göran Therborn, and Bob Jessop, Geoff Eley has questioned this interpretation of German "backwardness" and has sought to redefine the imperial German state in terms of its relative "autonomy" in relation to social groups, its particular organization as the "outcome of much wider political struggles," and its status as both an "institutional complex" and "permeable arena in which contending social and political forces interact."[41] This perspective has informed the important recent studies of George Steinmetz on the welfare state and Kathleen Canning on labor legislation in imperial Germany, both of which have moved decisively away from assumptions about the German state as a long-standing autonomous and homogeneous entity, toward analysis of its heterogeneous structure as an outcome of contingent historical processes.[42]

Recent theoretical discussion even more comprehensively challenges analysis of the state that presupposes its ontological essence as a "free-standing" power that is separate from other domains of social life and subject to various forms of seizure and deployment and that intervenes in society or economy from outside. Poulantzas defines the "institutional materiality" of any capitalist state as the outcome of a historically contingent and "specific material condensation of a relationship of forces among classes and class fractions," as a "strategic field and process of interconnecting power networks," including the organization of "processes of thought" in the context of state discourse.[43] In a more explicitly Foucaultian formulation, Timothy Mitchell argues that the presence of the state derives from the "complex of social practices" that constitute modern forms of "governmentality"—the dispersed political rationalities and disciplinary technologies that are deployed locally "around particular [social] issues," ranging from crime to public health to working-class radicalism, and institutionalized in prisons, clinics, schools, asylums, workhouses, welfare programs, and factories. From this perspective, the modern state emerges as a "structural effect," which appears to be a discrete entity removed from other realms of economic and social life. The boundary separating the state and other realms of economic and social life should be viewed not as a "boundary between two discrete entities but as a line drawn internally, within the network of institutional mechanisms through which a social and political order is maintained."[44] Consequently, the state is integral to the formation of workplace regimes, markets, modes of regulation and "governance," and regimes of capital accumulation that structure economic practices; it is not a freestanding body that intervenes in their "normal" workings from outside.[45] This cultural turn in state theory offers important insights for the study of political culture in the Saar region, where the state owned most of the coal mines, shaped the structures of industrial work by means of welfare and regulatory policy, and provided the legal frameworks for public political activity. It also allows new perspectives on the transition from paternalism to corporatism during the decade before 1914 and on the connections between state (re)formation and the development of corporatist ideological projects in Germany after 1914.

The Saar Region

The presence of the state, large-scale manufacturing, and a regional public sphere dominated by some of Germany's most powerful industrialists suggest the possibility of exploring developments in the Saar as a point of departure for understanding some of the wider social and political transformations of imperial Germany and early twentieth-century Europe. The Saar, one of Germany's leading industrial regions, witnessed dramatic increases in the size and output of its coal mining, iron and steel production, and glass manufacturing after 1870. Its concentration of large-scale industry, increasingly expanding unskilled and semiskilled labor forces, paternalist disciplinary and social policy regimes, and managerial and organizational changes after 1900 had parallels in other parts of Germany and Europe, but the political-cultural trajectory of employer politics in the Saar region during the prewar decade was also distinctive in its systematic elaboration and precocious development. In this way, it anticipated subsequent corporatist political-economic realignments in the rest of Germany and Europe during the early twentieth century.

The Saar Valley was once an isolated agrarian enclave, ruled by the Count of Nassau-Saarbrücken before the French invasion and occupation from 1792 to 1814. After the defeat of Napoleon, it was divided between Prussia and Bavaria, though the bulk of the coalfields went to Prussia. The Prussian state took over administration of nearly all Saar mines, which were placed under the direction of the Prussian Mining Office (Bergwerksdirektion). That office, located in Saarbrücken, was subordinated to the Upper Prussian Mining Administration (Oberbergamt) in Bonn and ultimately responsible to the Prussian Ministry of Public Works (Ministerium für öffentliche Arbeiten) in Berlin. If the mining industry nearly collapsed from the loss of French markets after 1815, the creation of the German Customs Union in 1834 opened up southern markets for Saar coal and led to a dramatic increase of coal output from the region. This occurred after the 1840s, when the Mining Office turned to a policy of labor recruitment that attracted increasing numbers of peasant migrants from the surrounding region, but especially after the 1850s and 1860s, when state officials introduced a series of economic reforms that removed restrictions on managerial prerogative and the "free" movement of labor, including the miners' traditional privileges and corporate or "estate" status.[46] In 1875, the Prussian Mining Office was the largest industrial enterprise in the region and employed nearly 23,000 miners. That figure rose to 30,376 in 1890 and to 42,418 in 1900. By 1910, its twelve "inspectorates" or collieries employed 53,055 miners. The Mining Office's production rose in similarly continuous fashion: in 1875, Saar mining was producing 4.5 million tons of coal annually; in 1900, just under 9.4 million tons; and in 1910, over 10.8 million tons.[47] By 1913, when its collieries were producing 18 million tons of coal, Saar mining constituted 8.6 percent of the coal industry in Germany.

The increasing output of the Saar mines took place within an expanding regional network of industrial concentration, involving especially the iron and steel mills and glass and ceramics foundries, with their demand for coke, as well as evolving regional markets in a range of industrial products. After French authorities privatized the iron and steel industry in the early nineteenth century, ownership of the major mills passed to the bourgeois Stumm family, originally from the Hunsrück region. These mills were located in Neunkirchen, Halberg, and Dillingen and employed only about 1,100 workers in 1846.[48] In 1852, the construction of a regional rail network made ore from Lorraine accessible to local factories and increased the efficiency of the shipment of iron products out of the Saar, enabling the subsequent expansion of the iron and steel industry. When Carl Ferdinand Stumm became a codirector of the firm Gebrüder Stumm in 1858, there was only one competitor in the region, the Burbach steelworks, under Belgian and Luxembourgeois ownership. But in 1873, another local industrial family, the Röchlings, built a steelworks in Völklingen, and by 1878, the Saar iron and steel industry employed 5,627 workers.[49] In subsequent decades, despite temporary downturns, the Saar iron and steel industry expanded dramatically: the number of workers employed by the five main concerns—the Stumm steelworks in Neunkirchen, the Röchling concerns in Völklingen, the Burbach steelworks in Malstatt-Burbach, the Brebach steelworks (formerly of Halberg), and the Dillingen steelworks—rose from 11,932 in 1890 to 17,830 in 1900 and 24,943 in 1910. Overall output of raw iron increased from 105,350 tons of raw iron in 1878 to 442,824 in 1890, 554,597 in 1900, and 1,197,688 in 1910. By 1913, the iron and steel mills in the Saar were producing 11.2 percent of all raw iron and 14.5 percent of all raw steel in Germany.[50] These developments were paralleled by the growth of a third major industry, glass and ceramics manufacturing. In the first decade of the nineteenth century, nearly all of the glass foundries were purchased by a handful of bourgeois families—the Vopelius, Reppert, Wentzel, Köhl, and Wagner families—and the ceramics industry was led by the Villeroy and Boch families, which joined their enterprises in Wallerfangen and Mettlach in 1842. By 1860, Saar glass foundries employed 1,500 workers; these figures rose to 2,500 in 1880 and 5,000 in 1906, when the largest concerns had workforces of some 250 employees. Its overall production figures never came close to the coal and steel industries, but the Saar glass industry was responsible for nearly one-quarter of all glass production in Germany by 1913.[51]

By the last third of the nineteenth century, heavy industry dominated the social landscape of the Saar Valley. The increasing pace and scale of industrial expansion and concentration produced larger and larger enterprises, and the major branches of Saar heavy industry employed increasing numbers of semiskilled and unskilled laborers. Consequently, the population of the Prussian Saar rose from 176,158 in 1843 to 652,294 in 1910.[52] It was concentrated in the urban industrial triangle bounded by Neunkirchen, Saarbrücken, and Völklingen and was disproportionately proletarian in relation to national averages. By 1909, Saarbrücken and St. Johann merged with Malstatt-Burbach to form a single large city [*Grossstadt*] with a population of just over one hundred thousand. Smaller industrial towns like Dudweiler and Malstatt-Burbach also witnessed rapid growth and highly mobile populations from the 1850s to 1914, though many *Saarländer* lived in small to medium-sized communities of between two thousand and thirty-two thousand inhabitants, and the region was home to a large number of so-called worker-peasants during this period.[53] Nevertheless, by the 1870s, few inhabitants of the region remained unaffected by the sprawling infrastructure of collieries, iron and steel concerns, metalworking plants, and glassmaking and ceramics foundries, as well as the communications networks (railroad, canals, and thoroughfares), municipal institutions and services, and communities that developed in their proximity to facilitate their operation.

These transformations helped to forge the distinctive political geography of the region, which was anchored in the local and regional dependencies of "company towns," notable politics, and the prominence of economic and political figures of national importance. The two leading political parties of the Saar, the National Liberal Party and the Free Conservative Party, were led by local industrialist notables from the iron and steel, glass and ceramics, machine-making, and coal industries and were challenged only by the Center Party after the 1890s. In this context, Saar industrialist notables attained levels of national preeminence and political influence that far exceeded the relatively small geographical size of the Saar after 1870: Stumm, a CVDI leader whose businesses extended to coal and ore fields and steelworks in Lorraine and Luxembourg, was a Free Conservative member of the Reichstag from 1867 to 1881 and from 1889 to 1901, a member of the Prussian Landtag from 1867 to 1870 and of the Prussian Upper House from 1882 to 1901, and a personal friend of the kaiser; Richard Vopelius was a Free Conservative member of the Prussian Landtag from 1876 to 1879 and from 1882 to 1903 and was chairman of the CVDI from 1904 to 1908; and Karl Röchling, owner of the Röchling steel concerns in Völklingen, was one of Germany's leading industrialists, with possessions in Lorraine, Aachen, and the Ruhr. As the leaders of regional employer associations and as national spokesmen for German heavy industry, Saar industrialists and their representatives were able to influence the formulation of state welfare and industrial policy during the 1870s and 1880s and to set the tone for much of the proposed antisocialist and antiunion legislation of the "Stumm era" during the middle to late 1890s.

After 1900, when new challenges from trade unions and social reformers brought on a crisis of paternalist hegemony and when local industrial enterprises entered into a process of technical-organizational transformation, a new generation of Saar employers and industry representatives of national prominence came to the fore. Alexander Tille, a former university professor and assistant to the chairman of the CVDI before he took over the Saar industry organizations in 1903, became the most prolific spokesmen for heavy industry in imperial Germany. Max Schlenker, Tille's replacement after 1912, became a leading industrial representative before the First World War and, after 1925, the managing director of the influential Langnamverein (Association for the Protection of the Common Economic Interests of the Rhineland and Westphalia), the main employer organization in the Ruhr. And Hermann Röchling, a son of Karl Röchling and technical director of the Röchling steelworks in Völklingen after 1901, became a leader of the national company union movement as member of the executive council of the German League of Company Unions (Bund Deutscher Werkvereine), leader of the German-Saarland People's Party (Deutsch-Saarländische Volkspartei) during the 1920s and early 1930s, and a leading Nazi industrialist after 1936. In these capacities, Saar employers and their representatives became the most systematic exponents of corporatist economic-social and political-constitutional change after 1908, up to and including attacks on universal male suffrage and the parliamentary system itself. They were thus crucial to the formulation of a new "fascist" ideological paradigm of industrial social order, with important consequences for the subsequent histories of industrial relations and political culture during the Weimar and Nazi eras.

CHAPTER I

Company Paternalism in the Industrial Saar

On 24 April 1890, workers of the Wendel coking plant in Hirschbach went on strike in an effort to secure wage increases, changes to the company pension and sickness fund, an eight-hour shift for workers at the blast furnace, and ten-hour shifts for all others. One week later, Dr. Hallwachs, the factory director, signaled his company's refusal to meet the demands or even to bargain with the workers: "A wage increase will not be granted. Whoever does not want to work can leave, but you will never work at this factory again. No Kaiser or king can help you here, for here we are the masters."[1] Hallwachs's dismissal of all authority beyond his own and that of his managers and his reference to the employers as "masters" of the factory were not products of the immediate circumstances of the labor dispute in Hirschbach during the spring of 1890. Rather, they expressed a set of understandings about employer prerogative and invoked a wider discourse about workplace relations that were common throughout the Saar and the industrial regions of Germany and Europe in the nineteenth century—a discourse that was variously institutionalized in the disciplinary codes, social provisions, and rituals that made up the paternalist factory regime in Saar mining, iron and steel production, and glass and ceramics manufacturing.[2]

This paternalist regime—the "Stumm system"—in Saar heavy industry is the subject of this chapter. German historians have interpreted this model as either a static preindustrial holdover or an entailed expression of underlying economic conditions and imperatives. This chapter examines paternalism as a historically determinate discourse about work relations and moral order, which was materialized in a conjuncturally specific set of factory institutions and managerial practices. It interprets the actual terms and propositions of paternalist discourse, its explicit claims and implicit assumptions, in order to demonstrate the ways it combined economically liberal definitions of the private labor contract and employer prerogative with moralizing and gendered understandings of work and the factory "family."[3] This chapter then examines the ways in which paternalist discourse was articulated to multiple forms of business ownership, a certain model of workplace discipline and company social welfare, a set of company rituals, and wider institutions of public authority that were entirely "rational" and appropriate to their times, particularly when they are compared with paternalist factory regimes in other regions of Germany and Europe.

In the final section, this chapter argues that the factory paternalism of Stumm and other Saar employers became a model for the imperial German welfare state of the 1870s to the 1880s. The latter differed from the other European welfare and regulatory measures in terms of its predominant focus on social insurance over state inspection and direct oversight, a divergence that was in large part due to the influence of Saar and other German industrialists in the making of the early welfare state. Rather than treating paternalism as an originary or self-contained managerial discourse derived from preindustrial values or from a managerial logic immanent to the monopoly phase of German capitalism, this chapter explores the ways in which the "Stumm system" was elaborated and sustained in historically contingent discursive-political fields and the specific institutional matrices of industrial organization and state formation in the latter half of the nineteenth century.

Saar Paternalism and Political Economy

Until very recently, the historiographical debate over industrial paternalism in Germany had been curiously silent about the actual terms and propositions of paternalism as a discourse and their articulations in managerial practices—that is, the complex of workplace discipline and social provision that constituted the paternalist factory regime. German historians committed to the preindustrial thesis have ignored Stumm's own self-description as a "modern man,"[4] while those concerned to show how paternalist ideas were compatible with the extralinguistic forces of monopoly capitalism have overlooked the productivity of paternalist discourse. But the economic terms and propositions of Saar paternalism were more than just responses to the evolving and autonomous relations and institutions of the industrial workplace in the monopoly phase of German capitalism; they were meaningful, productive, and highly mobile elements, subject to various kinds of articulatory practice during the imperial era. When read for their specific meanings in this way, it is clear that the economic dimensions of paternalist discourse in the Saar derived not from preindustrial or archaic values but from the implicit assumptions and overt commitments of bourgeois political economy and that they helped to shape the industrial workplace in the Saar.

Despite their presumptive invasiveness, Stumm often defended his managerial practices from the perspectives of bourgeois political economy and even rejected the "feudal" label given to him by his political opponents. In his numerous appearances from the 1860s on, before the Reichstag of the North German Confederation, the Prussian Upper House, the German Reichstag, local political meetings, and ceremonies with his own workers in Neunkirchen, Stumm described his work rules and welfare provisions in terms of bourgeois-liberal assumptions about the duties and responsibilities of the individual, the "free labor contract," and the economic imperatives and interests of modern industrial society.[5] He embraced the bourgeois ethos of self-determination and individualism, not only in his antitheoretical rejection of "book learning" (*Bücherweisheit*) in favor of a model of the employer who draws from the "world of practical life experience," but also in his description of the central dimension of the paternalist factory regime: its foundation in the "personal relationship between employer and worker."[6] The latter was understood in part as a relationship between two bargaining and self-interested individuals, partners in a negotiated "wage contract," in which employer and employee entered into an agreement over the terms and conditions of employment. This meant the employer's acceptance of workers as full "citizens" operating under the conditions of legal equality. In response to socialist claims that industrial workers had been "degraded to the level of a fourth estate," for example, Stumm insisted before his own workforce in 1895, "The working class [*Arbeiterschaft*] today is completely equal to all other categories of citizens, and I will never agree that the worker . . . possesses less value than a *Kommerzienrat* or a government minister." Moreover, he argued that because the hierarchy of positions in a factory was characterized by a wide range of job titles, tasks could not be divided into the two simple categories of employers and workers. "'Self-reliance is the name of the game' [*Selbst ist der Mann*] is the principle of every diligent worker," Stumm maintained, arguing that this principle would get the worker "the furthest in relation to the employer."[7] This view corresponded with an understanding that limited the role of the state in the industrial workplace and that bore some traces of the "night watchman state" of economic liberalism. During the Reichstag debate over the new Industrial Code Bill in 1890, Stumm explicitly defined the state as a guarantor of the "freedom of the wage contract"; its role was to defend the allegedly natural relations of authority that obtained in civil society.

> In my view the state should not intervene directly or in a partisan way in these relationships, and at the very least it may not undermine authority where authority is present. The state must value and protect the authority of the employer just as much as it does the authority of its own organs, and those of the churches and the schools. Naturally, I find it completely appropriate that the state enacts measures in order to prevent the misuse of this authority; for the misuse of authority damages authority in general.[8]

Stumm therefore objected to attempts by the state to restrict employer control over employee behavior outside of the factory and to regulate the degree of penalties employers could impose on workers as well as factory work rules in general: "the legislator has nothing to say about the content of the factory work code . . . that must be left to the agreement between worker and employer." Such interference violated a "fundamental principle" of the "free labor contract": the right of workers and employers to agree on the terms of employment freely and independently.[9] Like many nineteenth-century political economists and despite his willingness routinely to call on the state to crush independent labor organizations, Stumm ignored the glaring asymmetry of power and conditions of existence between the wealthy industrialist and the wage earner, whose reliance on paid employment for survival could not simply be reduced to a "choice" between entering into a "wage contract" or not.

These claims were deemed compatible with the most far-reaching disciplinary rules of the paternalist factory regime. It was on the basis of the need for reliable, diligent labor in the factory that Stumm defended his practice of penalizing workers at his steelworks (especially by means of workplace fines) for civil infractions taking place outside of work. In 1891, he insisted that "the behavior of workers in and outside of the factory absolutely cannot be separated from each other."

> A worker who dedicates himself to dissolute moral conduct outside the factory will also not be able to function at the workplace. If he or a foreman runs a store or a business, then connections will necessarily be established between foreman and workers that would have some influence on their relationship in the factory to the disadvantage of other workers. If workers frivolously file charges in court, say, because their wives have insulted one another, then it will be impossible to prevent the conflict and quarrel from entering the factory. If half-ripe lads, who do not earn sufficient wages, get married prematurely and bring children into the world, then they will not be in a position to support and rear the latter, and they will lose the necessary strength and enthusiasm to do their work. Whoever does not followed the principle "honesty is the best policy" in his private life will not be able to resist the temptation to appropriate valuable materials, which lie about unlocked in the factory.[10]

In this sense, Stumm understood his factory work rules and disciplinary regime as a means "to develop" each employee into "a competent and well-behaved worker."[11] This kind of discipline, Stumm argued in 1898, was "absolutely necessary in a rational enterprise that is supposed to remain competitive."[12] Moreover, his unabashed commitment to excluding trade unions from his steelworks and attacking local publicans even for making their rooms available to trade unions in Neunkirchen derived from his rejection of collective negotiations of any kind. The association of workers, in that sense, violated the "freedom" of the individual and subverted the "personal relationship" between employer and employee. Stumm vehemently opposed social democracy not only for its revolutionary doctrine and its variously "immoral" commitments—including its opposition to the bourgeois family and marriage, the churches, the monarchy, and state authority—but also for the ways its supporters brought their "tyranny" to workers.[13] In a curious rhetorical inversion, he claimed that the real threat to the "freedom" of workers was not excessive paternalist control over workers' lives, anchored in the asymmetry of social power between employer and employee, but the coercive threats to a worker's "body and life" in the efforts of independent trade union "agitators" and Social Democratic "seducers of workers" to organize and mobilize workers for industrial actions and collective bargaining. When a worker joined such a "combative organization," he was brought under the spell of "outside agitators" who "destroyed his independence."[14] By 1898, Stumm was celebrating "American industry"—the classic signifier of the economic modern for contemporaries and subsequent scholars—for its efficiency, productivity, and growing influence over world markets at the expense of the British; he attributed this superiority mainly to the fact that American employers had achieved "mastery" over the trade unions.[15] Finally, as Stumm maintained in his speeches before the Reichstag in 1890, paternalist control over the lives of workers off the job, including the infamous marriage clause at the Neunkirchen steelworks, was authorized not by "feudal" privilege or birthright but in part by the "free wage contract" itself, which contained these rules and allowed the worker to accept them or not. This, according to Stumm, was a coercion-free exchange: "If a worker does not comply with this arrangement [i.e., the marriage clause], he will not be penalized; rather, I would pose a question like: do you want to submit to the rules that I have presented to you or do you want to give up this job? This has nothing to do with punishment."[16]

The social provisions associated with the paternalist factory regime, from bonuses to company housing, were framed by a similar capitalist logic. Stumm and other Saar industrialists never made a secret of the "rational" economic motivations behind their extensive benefits and welfare programs. In their main press organ, the *Saarbrücker Gewerbeblatt,* they described the "patriarchal . . . welfare provisions" of local heavy industry as a means for promoting the "good behavior" of workers and their "long-term competence" and for reducing labor turnover and accidents at the workplace—all of which contributed to the "prosperity of the . . . factory."[17] In a similar way, Stumm described the importance of wage bonuses in securing a "lasting, settled population" in order to reduce labor mobility and to stabilize his labor force.[18] This was a common motivation behind social provisions, echoed in the statements of officials at the foreign-owned Burbach steelworks and of Saar mining officials, who referred to the "purpose" of attracting a "core of workers" (*Arbeiterstamm*).[19] Company housing and housing loans and premiums, which extended employer prerogative most comprehensively into the everyday private lives of workers and their families, were similarly understood as mechanisms for securing a stable workforce and for inculcating the habits of "economic man." According to an essay cited in the *Saarbrücker Gewerbeblatt,* a worker's "property makes him thrifty and economical [*haushälter-*

isch]," so that "he learns to value and put to use the smallest savings for his household."[20]

The Moral Dimensions of Work and the Factory "Family"

Saar paternalism articulated other meanings that cannot be reduced to these strictly economic concerns and ambitions. The paternalist factory regime was also structured in a relation of moral tutelage in which work was defined by its moral properties and in which the employer or factory director was figured as the moral guardian of workers. In this context, comprehensive work rules and welfare programs were explicitly motivated by moral concerns and designed to attend to the moral development of workers. Moreover, these intentions and practices were framed by a wider paternalist vision of a factory "family," ordered according to bourgeois codes of sexual difference, and instantiated in the regime of discipline and social provision. In these ways, Saar paternalism combined capitalist rationalities with bourgeois concerns about moral and familial order in an ideological discourse capable of interarticulating diverse—though not incompatible—assumptions and intentions.

Saar employers routinely invoked morality and moral concerns in their dealings with workers and wrote moral concerns directly into their factory work rules, which were designed to inculcate in workers what local employers defined as respectable and moral behavior. Stumm explicitly described his relationship to his workers as one of moral tutelage, motivated by his "moral sense of duty" and "Christian conviction." He believed that the moral "education of the worker" by means of his disciplinary code, one of the "moral obligations" prescribed by his "conscience" and his "God," was more important than the social provisions of the paternalist regime because it had the most direct effect on the behavior of his workers.[21] Other employers shared this perspective. Toward this end, several local firms promulgated work rules that targeted employee behavior both inside and outside of the factory. Stumm announced bluntly in work rule 44, "Every foreman and worker will conduct himself outside the factory in a way which corresponds to the dignity of the house of *Gebrüder* Stumm; you can expect that your private behavior will always be subject to the firm's scrutiny, and that improper conduct outside the factory will result in termination when there is no applicable penalty in the existing factory regulations."[22] At most large firms, workers were subjected to fines or dismissal for drinking, sleeping, or violation of factory discipline, but they were also punished for "violations against good morals" and activities related to civil "disturbances" and "excesses" (e.g., public drunkenness; the discharge of weapons during Christening celebrations, weddings, or New Year's parties; or resistance against police officials).[23] Officials of the Mining Office commanded their employees to behave in such a way as "to bring honor to your estate and occupation," especially in relation their workplace superiors, and penalized miners for various kinds of civil infractions or other "excesses," including fighting that took place away from the collieries.[24] Company officials at private Saar firms also monitored other aspects of the moral lives of their workforce. At the Stumm concerns in Neunkirchen, foremen were required to observe the drinking habits of their employees: those below forty years of age who "appeared" to drink too much and thus represented both a moral threat and a future financial burden on the company's sickness fund were immediately dismissed. The disciplinary code demanded that they report all manner of employee "excess" taking place "outside of the factory."[25]

Moreover, the rules about moral behavior, obedience, discipline, and honor were generally enforced in all company facilities, including cafeterias, washing rooms, hospitals, the schools, swimming pools, parks, and dormitories.[26] In the case of Saar mining, the latter, which were led by wardens or "house masters" and policed by "resident elders" (*Stubenältesten*), enforced strict obedience to a wide array of disciplinary and moralizing rules: proscribed behavior included violating the 9:30 p.m. curfew, failing to be in bed by 10:00 p.m., making loud noises, engaging in disputes, singing "indecent songs," and even whistling. All residents were forced to adhere to dress codes mandating appropriate attire—all were required to wear shirts, pants, and shoes when walking around the facilities—and, not surprisingly, access to women was carefully regulated. Fears of illicit sexual behavior and its moral effects on the workers meant that women were only allowed to enter the dormitories during the daytime for the purpose of bringing food or clothing to residents.[27] In general, a concern with "morals" and respectability, in addition to workplace obedience and subordination, inspired many of the most imperious work rules, including Stumm's rule about workers greeting him and the similar Mining Office rule about residents of the mining dormitories standing and removing their caps when their superiors entered the rooms or the areas surrounding the facilities.[28]

Similar kinds of moral considerations motivated the many social benefits and programs of the paternalist factory regime. Officials of the Mining Office understood the function of the miners' *Knappschaft* [sickness and pension fund] especially its sponsorship of schools for children, in terms of its role in bringing about the "moralization" of miners and their families, and their housing and settlement policies were similarly conceived as attempts to "elevate the economic as well as the moral well-being of the worker" into an "orderly" and "regulated lifestyle" in order to prevent the formation of a "massive and morally and politically dangerous proletariat."[29] Stumm routinely invoked the "moral" benefits associated with his social provisions and turned them into forms of moral discipline: they were available only to employees who demonstrated the appropriate moral comportment; workers who violated the standards set by the firm or in general failed to maintain a "thoroughly proper lifestyle" were denied all bonuses, favors, or other benefits.[30] Employers generally considered these ambitions or intentions entirely compatible with, indeed essential to, "rational" managerial practices, since diligent labor was predicated on good "morals."

This moralizing framework reflected a bourgeois, not a "feudal," orientation toward values and behavior. The kinds of "morals" that Saar employers hoped to inculcate in their workers—thrift, sobriety, moderation, self-reliance, self-discipline, sense of duty, loyalty, sexual restraint and propriety, respectability, family life, "good citizenship," and especially a positive attitude toward work—were all part of the ensemble of virtues and social practices associated with *Bürgerlichkeit,* or "bourgeois culture," the focus of much social-historical research on the German middle classes [*Bürgertum*] in the wake of the *Sonderweg* debates of the 1980s.[31] When René von Boch spoke to his workers in 1891 during the fiftieth anniversary celebration of his ceramics firm in Mettlach, for example, he talked about the exemplary qualities of past employees, emphasizing the "fear of God, fulfillment of one's duty, love of work, mutual trust" as the Villeroy & Boch "solution."[32] Later that same year, Hans Seebohm, director of the Burbach steelworks, also invoked bourgeois values during a commemoration ceremony. Congratulating the recipients of pocket watches for their twenty-five years of service to the firm, he emphasized their "diligence and skill" but "above all" their "moral way of life" (*sittlichen Lebenswandel*).[33] Similarly, Stumm routinely congratulated his workers for their "good behavior, diligence, and competent performance" on the job.[34] Even his references to the military and military discipline, which some historians have interpreted as "feudal" gestures, are best understood in terms of the values they were meant to celebrate. When he referred to the army as a "school for life" during a ceremony for the local regiment in 1896, he meant to point up its role in inculcating the virtues of "manly discipline, loyalty to the monarch, patriotism, sense of duty, and willingness to sacrifice," which the soldier "brings into bourgeois life, in the family."[35] In other words, Stumm and other Saar employers were preoccupied with the bourgeois virtues of work and home life, rather than military virtues per se. In this regard, Stumm's primary concern was to elevate workers to the ranks of middle-class citizens (*Bürger*) in order to sustain the independent *Mittelstand,* that vital "link between the propertyless and the rich."[36] From this perspective, the regime of disciplinary rules and social provisions of Saar paternalism was understood as a way of treating the worker not "simply like a work machine" but as a "decent and moral" person in order to "maintain public morals"—the foundations of all "bourgeois and state order."[37]

In this bourgeois framework of assumptions and meanings, Saar employers associated work with moral qualities and attributes. They celebrated hard work and labor discipline for their moralizing aspects: work was considered a "means of improvement" that provided for an individual's material needs; inculcated "values," such as orderliness and moderation; and deterred the individual from engaging in "idle" or dissolute pursuits.[38] Its opposite, "laziness," was to be condemned or even criminalized. Reflecting on the "sacredness" of work and the imperative to "maintain the desire to work," the industry friendly *Malstatt-Burbacher Zeitung,* suggested, "Whoever avoids [this] most holy earthly undertaking, whoever thereby contributes to the dissolution of all bonds of honor and morals and the subversion of public order and morality, does not deserve the slightest consideration or humane leniency; rather, the full force of the law should be applied to such people."[39] In this sense, Saar employer-paternalists drew on the long-standing bourgeois representation of labor as an honorable human pursuit, an ennobling activity that endows its practitioner with both moral and political significance.[40] During firm ceremonies and dinners for the distribution of bonuses and the induction of employees into sickness and pension funds, Saar employers routinely invoked the "honorable" and moral aspects of diligent work and long-term service to the company. Conversely, as discussed previously, Saar employers insisted that what they deemed to be moral rectitude was the only solid basis for steady, diligent labor. This was why they made few distinctions between discipline and orderliness in the factory and moral order more generally:

Stumm maintained, "A worker who dedicates himself to dissolute moral conduct outside the factory will also not be able to function at the workplace."[41] In this sense, Saar industrialists resisted clear distinctions between morals, on the one hand, and strictly economic considerations, on the other.

The most obviously bourgeois aspect of the paternalist factory regime, however, was its gendered vision of familial order, which served as the model for labor relations throughout the Saar from the 1860s to the 1890s. The paternalist factory regime was discursively founded on a set of pervasive assumptions about work relations that identified the employer as the "provider" (*Brotgeber*) or father figure while it identified the workers as the dependents or children in the larger "family of producers" (*Arbeiterfamilie*). The Free Conservative Stumm, for example, regularly referred to himself as the "head of the larger Neunkirchen family of workers" in his role as owner of his steel factory in Neunkirchen; and the National Liberal and Catholic René von Boch of the Villeroy & Boch ceramics foundry praised his long-standing and "excellent family relationship" with his workers in Mettlach.[42] Derived from the model of the bourgeois family, this gendered imagery served to define relations of power in the factory: the employer, as the male head of the factory family, was responsible for all decisions related to the terms and conditions of work. He claimed the right to establish wage rates, work hours, and extensive work rules without the input of workers. Similarly, he rejected independent trade unions or workers' representatives as threats to the sanctity of private property—an owner's right to dispose of his property as he saw fit—and to the employer's authority as "master of the house" (*Herr im Hause*) or factory. Yet paternalist work relations were not based on total domination over workers; they rested instead on an implicit (and often explicit) understanding of reciprocal obligation and duty. Stumm explained, "The employer should feel as though he were the head of a large family, whose individual members have a claim on his munificence so long as they prove themselves worthy."[43]

This was a comprehensive vision of industrial and social order, anchored initially in the family-owned firm involving both the factory owner and his spouse. Well into the 1890s, Saar industrialists supported the family-owned firm as the best model for industrial organization. In an 1888 petition to the Prussian Ministry of Commerce, leaders of the Saar Chamber of Commerce supported legislation in favor of a new kind of "public company with limited liability" that would maintain the organization of the "family firm" even when no members of the "founding family" were interested in the actual running of the company. According to the petition, allowing family members to participate by means of the ownership of shares in the company had clear "economic and political advantages."

> The enterprise itself will continue to exist uninterrupted in the same old ways. The shareholders, who are closely connected, will endeavor to survive with great energy, even in the face of losses during bad times, certainly with more vigor than just any other firm with fixed investments. Keeping the connections between the family spirit and enduring enterprises will result in each elevating and strengthening the other. The favorable consequences for the formation of a permanent core of workers, loyal to the firm and the family, will necessarily follow.[44]

In this sense, the factory "family" linked the actual familial basis of firm ownership to relations between employers and workers; it thereby involved both the employer as father figure and the employer's wife as "mother" in the regime of moral discipline and welfare that characterized Saar paternalism. As the *Saarbrücker Gewerbeblatt* noted, "all of these programs" and facilities were supposed to be administered "under the constant personal oversight of the boss of the firm and his spouse."[45] In the Saar, therefore, the wives of leading industrialists took part in a range of charitable activities associated with the paternalist factory regime, including overseeing the factory schools for girls, child care initiatives, charitable distributions, and Christmas events and activities. As the next chapter will suggest, this activity was sponsored by the firm and linked to the local women's association (*Frauenverein*), which was usually run by the wife of a local industrialist.

Indeed, factory rituals and ceremonial provided important occasions for the enactment of the "familial" relationship in the factory. Saar employers extended their assumptions about the comfortable intimacy between members of the family unit to their relations with workers; this was part and parcel of their insistence on the fundamentally "personal" aspect of all work relations. Regular and intimate employer-employee contacts, institutionalized in the employer's weekly office hours or in the familiar setting of factory rituals and ceremonies, were designed to facilitate a harmonious "work relationship" and mutual respect and affection between employer and employees—respect and affection that were enforced by means of the factory work rules, which required workers to greet management by standing and removing their caps.[46] Saar firms sponsored a busy schedule of factory events and rituals during which such familial relations were regularly reinforced: these included the aforementioned ceremonies for the distribution of bonuses and the induction of members into the company-sponsored sickness and pension funds; lavishly orchestrated visits by local and national dignitaries; holiday and patriotic celebrations; and, most symbolically, festivities that marked important events in the life of the factory owner and his own family (e.g., birthdays, weddings, and anniversaries).[47] Anniversaries marking a firm's twenty-fifth or fiftieth year of operation commonly involved owners, managers, workers, and their families. Such ceremonies often began with a midday meal (*Festessen*) to which the owners, company officials, and a few select workers were invited. Typically, speeches in honor of the owners and in praise of a loyal workforce would be followed by toasts. The celebration would broaden afterward to include all workers and their families, gathered in open areas (e.g., forest areas or the nearby mining picnic grounds) where they were provided food, drink, and various forms of entertainment. The latter usually involved music performed by the company-sponsored choral societies and musical bands as well as dancing, carousels, show booths, and prize contests. The festivities often involved the broader community; at the fifty-year anniversary celebration of the Raspiller & Co. glassworks in Fenne in 1884, for example, both the factory and town buildings were decorated with flags, garlands, and wreaths during the day's festivities. Local clubs and associations also participated in the evening festivities.[48] Similarly, during the fifty-year anniversary celebration of the firm Villeroy & Boch, which included managers, workers, and other invited guests in a series of ceremonies and events, Boch narrated the history of the "family" firm and invoked the familial relations between employer and workers.[49] These kinds of "family" activities complemented the more direct ceremonies associated with the major life events of the employer's family—for example, Theodor Sehmer's fiftieth birthday celebration in 1897, which included employees and the singing club of his machine factory, the choir from the local regiment, a Chinese lantern procession and serenade, and a dinner in a local restaurant; or the wedding reception held for Stumm's daughter, which included workers of his steelworks in a series of festivities, including a concert, in his park surrounding his estate.[50]

As this gendered activity suggests, the bourgeois model of the family authorized a range of tutelary practices and work rules that involved the regulation of the private sexual and marital lives of workers and their families. In Neunkirchen, young workers could be sacked for illicit (i.e., nonmarital) cohabitation, and at several firms, workers who wished to marry were expected to seek approval for their marriage from their employer. In the case of marriage, workers had to prove they were of the requisite age and possessed the necessary financial means and that the prospective wife demonstrated the appropriate moral virtue. This practice was not motivated by "feudal" ambitions; it was a consummately bourgeois practice. As Stumm explained, it was an attempt to address a problem widely discussed in middle-class circles: the "breakdown of the working-class family."

> When a worker marries in his early years, before he has earned a high wage and before he can save enough, all collapses at the first downturn of the economy. If the man becomes ill or is temporarily unemployed, immediately difficulties arise, which often lead to drinking, dishonesty, or other ruinous consequences. They grow in proportion as the size of the family increases, and the complaints raised about them in some heavily populated communities are well known. Many such marriages are conceived in the delirium [*Taumel*] of the *Kirmes* [religious ceremony] or in the greatest wantonness [*Leichtfertigkeit*]—and in a few years bitterly regretted.[51]

By requiring approval before marriage, Stumm felt he could ensure the success of the proposed union—a concern shared by other Saar factory owners and directors of diverse political affiliations.[52]

The paternalist preoccupation with morality, orderly conduct, and sexual and marital relations among workers also found expression in the aims and practices of company social provision. The most important goal of this kind of company munificence was the cultivation of orderly family arrangements, conceived according to a bourgeois model of moral comportment and family life. This family ideal was to be realized in a dual sense. First, company welfare served to foster "friendly, patriarchal relations between employer and employees" and institutionalized the relations of the factory family, by designating the employer as the provider while designating the workers as recipients of social provision.[53] This was a relationship of hierarchy that was routinely reinforced during factory ceremonies, which brought employers together with workers for the presentation of company benevolence and largesse, and in various communications between factory management and the workforce. For example, when, during a factory ceremony in 1890, Director Hans Seebohm of the Burbach steelworks warned his employees not to "allow themselves to be misled into their own ruin" by joining an independent labor organization, he was reemphasizing the role of the firm as "provider" of workers and their families.[54]

Second, the benefits offered by large industrial firms in the Saar were designed to cultivate the bourgeois family norm among workers or within working-class communities by defining and sustaining the male worker as the principal breadwinner and his spouse principally as wife and mother. In Saar mining, this was the explicit motivation behind housing premiums, which were offered only to workers over twenty-five years of age with families. The male breadwinner norm was also structured into the sickness and pension funds, which calculated and distributed their benefits on the basis of a worker's familial position. Indeed, the settlement policies of the Mining Office, which aimed to avoid concentrated proletarian rental housing, were framed by a general bourgeois logic of family living. The latter involved smaller, single-family homes with small gardens that allowed for the "most advantageous economic and moral" employment for miners' wives, who could then serve as both caretakers of children and agricultural cultivators and thus contribute to the well-being of the family.[55] In the Saar steel industry, many benefits were also only distributed to male workers (women were not hired by the local steel mills) who were deemed of the appropriate age and disposition and who could demonstrate their willingness to assume the responsibilities of establishing a household. At the Stumm steelworks, workers' spouses were limited to the role of housewife and mother by the factory code, according to which spouses were not allowed to seek employment with other firms or to establish their own businesses. Moreover, Stumm offered housing, housing loans, and other material incentives only to those workers who complied with his marriage and other morals clauses.[56] In a similar way, owners of the Vopelius glassworks openly acknowledged the gendered motivations behind their social provisions. As with the sickness and pension funds in mining and iron and steel production, the Vopelius glassworks distributed payments according to a worker's familial position, and its other programs, including factory schools for children and girls, were designed explicitly to enhance the "family life" of workers, to allow women to attend to their "motherly duties" in the home, and to promote a "healthy domesticity"—all of which constituted a "bulwark against the Social Democratic movement."[57]

In these ways, Saar employers attempted to regulate the marital relations and to shape the family life of their employees. The moralizing ideal held out to workers in the Saar by employers generally was the single-family home, that discrete social space organized around rigid lines of sexual difference and removed from the proletarianizing effects and "public" dimensions—the street life, community networks, and multiple family apartments—of workers' settlements. This ideal was implemented differently and with different degrees of success in the Saar. The Mining Office relied on dormitories, in part to allow miners to maintain their rural holdings, and the Burbach steelworks relied on urban workers' settlements to a much greater extent than Stumm, who tried to offer his workers the opportunity to purchase single-family homes. Yet even Stumm was forced to build workers' dormitories in Neunkirchen, and all Saar employers, including officials at the Burbach steelworks, embraced the ideal of the single-family home. They considered the latter, with husband, wife, and children, as the best means of inducting workers into the ways of a "thoroughly proper lifestyle" on the basis of moral probity, "stable" family relations, and bourgeois gender order.[58]

Saar, German, and European Paternalisms

The model of paternalist work relations known as the "Stumm system" was widely shared in the Saar, in part because it was coordinated and enforced regionally, under Stumm's leadership, by means of common discussion and efforts in the leading business organizations, the Saar Chamber of Commerce, and the local Prussian Mining Office. It was secured by effective regionwide blacklists and union-breaking measures when such tactics were deemed necessary. Deployed against socialists who attempted to organize in the region in 1877, such tactics were subsequently reinforced during the Saar miners' strikes from 1889 to 1893 and during the Christian and socialist union drives of 1903–4. These efforts involved consultation within the employers' organizations; coordinated announcements and policy statements in the central employer press organ, the *Saarbrücker Gewerbeblatt,* as well as industry-friendly newspapers (especially the *Saarbrücker Zeitung*); and deliberations between local state officials— including mayors and the *Landräte* [county commissioners] and Saar employers.

Despite this regionwide cooperation and these shared concerns, however, it is important to note the different articulations of paternalism and the religious and political diversity of paternalist employers, even in the Saar region. The ensemble of disciplinary rules and social provisions of the paternalist regime was embraced by different employers and compatible with different forms of business ownership. It was implemented by Stumm, a Calvinist and Free Conservative, at his steelworks in Neunkirchen, Brebach, and Dillingen; by Karl Röchling, a Protestant National Liberal whose conflict with Stumm over economic policies in relation to foreign trade are often cited as proof of his liberalism, at his steelworks in Völklingen; by Hans Seebohm, the Protestant National Liberal director of the Burbach steelworks, which was not family owned but part of a foreign-owned joint-stock company, the Luxemburger Bergwerks- und Saarbrücker Eisenhüttengesellschaft; and by René von Boch, a Catholic National Liberal, at his ceramics concerns in Mettlach and Wallerfangen. This ensemble included various benefits, from wage bonuses to company housing, but also invasive factory work rules like the marriage clause, which was not confined to the Stumm steelworks in Neunkirchen; there were similar clauses, for example, at the Burbach steelworks and the ceramics foundries of Villeroy & Boch.[59] Moreover, some of the classic signs of what was once thought to be a "feudal" entrepreneurial class were broadly shared across the region. In this regard, it is telling that not only Stumm but Röchling, Villeroy, the leather manufacturer Heinrich Korn of Saarbrücken, and even Theodor Sehmer, the less antiunion, Protestant-liberal owner of the Ehrhart & Sehmer machine-making factory in Schleifmühle, lived in "villas" rather than modest homes. In their architectural styles and apportioning of space, the villas were designed as "expressions of the representational needs" of a self-confident bourgeois class, "affluent *Bürger* and manufacturers," not as representations of feudal relics of a bygone era.[60]

The paternalist practices of Sehmer serve to highlight both the diversity of paternalisms and the way in which one version, the "Stumm system," largely succeeded in imposing its own dominance in the region. Sehmer, who built his factory with Ludwig Ehrhart in 1876, adopted many of the paternalist institutions and practices of other Saar employers: he created a support and pension fund for his workers as early as 1876; the factory had its own choir association (*Gesangverein*); and employees participated in factory ceremonies and rituals, during which Sehmer extolled the virtues of the "peaceful" relations between employer and employee.[61]

But Sehmer frustrated local state officials and other industrialists with his relative toleration of (or at least failure to police) workers with socialist sympathies. During the antisocialist period of the mid-1880s, a number of workers at Ehrhart & Sehmer, along with workers at the state-owned rail yards in St. Johann, were discovered to be Social Democrats. Local officials warned Sehmer, who opposed social democracy, but he disputed the socialist affiliations of his workers and did not take action against them. Only in mid-October 1884—when state officials, including the mayor of Malstatt-Burbach, the *Landrat* in Saarbrücken, and the *Regierungspräsident* [District Governor] in Trier, joined with local employers, including especially Stumm himself, and approached Sehmer—did he dismiss the socialists among his workers.[62] This amounted to a systematic campaign, in which local police and state officials searched the homes of suspected Social Democrats, arrested them, and even attempted to force the principal "agitators" out of the region. At least until the early 1890s, it effectively ended any serious Social Democratic activity in the region.

Moreover, the range of paternalist factory regimes (their varied articulations in relation to forms of business organization, industry, religious affiliation, and party political orientation) are evident from broader German and European comparisons—an indication of the wider resonance and productivity of paternalist vocabulary and practice obscured by the single-minded focus on Krupp and Stumm in German historiography.[63] In Germany, invocations of the factory family and the employer as provider, the stress on personal relations between "master" and "man," and efforts to institute extensive work rules and company benefits and to orchestrate factory ceremonies in order to secure a stable labor force and "harmonious" work relations were not limited to the large-scale mining and iron- and steel-producing firms associated with the CVDI or the Ruhr-dominated Association of Iron and Steel Industrialists (Verein Deutscher Eisen- und Stahlindustrieller, hereafter VDESI).[64] They also variously defined the labor regimes of large textile firms in the Rhineland and Westphalia, the electrotechnical and "rationalizing" firms like Siemens, chemical concerns like Bayer in Elberfeld and Leverkusen, or the Hohner musical instrument factory in Württemberg. As Thomas Welskopp reminds us, they were integral to the managerial practices of a liberal manufacturer like Friedrich Harkort, whose machine factory in Wetter in the Ruhr was one of the first firms to develop a paternalist system. In the case of Harkort, "company patriarchalism, market radicalism, and liberal engagement went entirely hand-in-hand."[65] Moreover, paternalist discourses and institutions, which were politically and religiously diverse, were evident in a variety of industries across Europe during the nineteenth century: they shaped labor relations in mining in the northeast of England and Scotland and northern and southeastern France; steelmaking in northeast and northwest England and northern and northeastern France; and textile manufacturing in Lancashire and the West Riding in England and in Roubaix, Rouen, and Alsace in France.[66]

Saar Paternalism and State *Sozialpolitik*

If Saar paternalism bore similarities with paternalisms in other regions and other parts of Europe, the distinctiveness of the German case lies in the capacity of heavy industrialists to impose much of their version of paternalism on the nation as a whole during the 1870s and 1880s. As a model for industrial relations, the paternalist regime in the Saar, Ruhr, and other centers of German heavy industry was extremely influential in shaping the new state social policies introduced under Bismarck direction by the imperial government in the 1880s, even if state officials like Theodor Lohmann of the Prussian Ministry of Commerce actually drafted key welfare policies.[67] In this way, work relations and workplace institutions in the Saar were closely interarticulated with state regulatory and welfare institutions in ways that secured the paternalist model both at the level of state welfare and at the industrial workplace during the 1870s and 1880s.

The first comprehensive response to socialist activity in the Saar came not from the Prussian or imperial governments in Berlin but from local employers and state officials after the Reichstag elections of January 1877. Initially, Social Democrats attempted to organize Saar workers briefly in 1872, but they were thwarted by local resistance and state repression. During the more sustained recruitment drive of March and May 1876, Social Democratic meetings in St. Johann, which were coordinated out of Mannheim, were also shut down by local police. Social Democratic organizers Carl Hackenberger and Friedrich Wilhelm Raspe, who studiously refrained from making controversial statements, were promptly sentenced in April and May to prison terms of one year and one month, respectively, for promoting the cause of the party. The turning point for Saar industrialists came in the summer of 1877, when Social Democrats Hackenberger, now released from prison, and Harry Kaulitz began to hold meetings "almost day by day in the most disparate localities," including the mining communities, and began publishing their own newspaper, *Die Freie Volksstimme,* in St. Johann. Convinced that the efforts of mining officials to confront Social Democratic speakers during their meetings would not keep the movement in check, Mining Office chairman Adolf Aschenbach invited the employers of Saar heavy industry to a meeting on 6 June, during which they agreed to the formation of the Employers' Committee for the Struggle against Social Democracy.[68] Under the leadership of Aschenbach and Stumm, the committee declared a ban on their employees' "direct and indirect" involvement in Social Democratic activities, especially the reading or distribution of party-related literature, taking part in socialist meetings or associations, or visiting "taverns in which Social Democratic meetings are held or literature of this movement is available." Any worker who violated the decree would no longer find work at any of the participating firms, which in June employed some thirty-six thousand workers.[69]

The committee instituted a regionwide network of surveillance and blacklisting, its own "Anti-Socialist Law,"[70] which effectively suppressed the Social Democratic movement in the Saar. It sent its own officials, "reliable persons," and even a stenographer to Social Democratic meetings in order to gather information about which workers were attending and to record the statements of meeting organizers and participants. The reports were shared among the firms and with local state officials, mayors, and the *Landrat* in Saarbrücken, so that disciplinary and legal action could be taken against their employees and socialist activists. Local authorities acted swiftly and ruthlessly. By the end of July 1877, Hackenberger and Kaulitz were under arrest and charged with "incitement" to "class hatred" and "resistance against the authority of the state"; in August, they were both sentenced to two and one-half years in prison. Kaulitz's three replacements as editor of *Die Freie Volksstimme* were all arrested, one after another, in July and August, and the newspaper's offices were sealed; Kaulitz's bookstore was closed, and its contents were confiscated in October; and the newspaper's colporteur, Franz Sater, was expelled from the region in November. Four other socialist organizers escaped to Belgium by August, while the last remaining activist, Franz Heinrich Mathis, was arrested in December for selling Social Democratic brochures.[71] In 1881, the Mining Office, a "neutral" state institution, withdrew from the committee, but Saar employers maintained the surveillance and blacklisting practices under the auspices of their new regional organizations, the Southwest Group of the Association of Iron and Steel Industrialists (Südwestliche Gruppe des Vereins Deutscher Eisen- und Stahlindustrieller, hereafter SGVDESI) and the Association for the Protection of the Common Economic Interests of the Saar Region (Verein zur Wahrung der gemeinsamen wirtschaftlichen Interessen der Saarindustrie, hereafter VWGWISI), established in the summer of 1882.[72]

This regional coordination of antisocialist and antiunion suppression anticipated similar measures implemented in other industrial regions and the Anti-Socialist Law introduced in the Reichstag by the imperial government the following year. In the context of regional Social Democratic mobilization, employers in the Ruhr met in mid-June 1878 to plan their own antisocialist measures. When some six hundred industrialists met in Düsseldorf toward the end of June in a large meeting sponsored by the Ruhr-based Langnamverein, they announced a regionwide effort to employ "all legally and morally permissible means" to silence Social Democratic activity among industrial workers.[73] Similar kinds of actions followed in other industrial areas—in the region around Dortmund and in the districts of Solingen, Lennep (especially in the cities of Ronsdorf and Remscheid), Neuwied, Cologne, Koblenz, Kreuznach, and Wetzlar. In many cases, these actions were undertaken by employers in large-scale industry in conjunction with local state officials.[74] Moreover, the latter often recognized these actions and held them up as models for state action, just as employers attempted to prompt state officials to undertake similar kinds of measures. Indeed, the meeting of Ruhr industrialists in late June 1878 was designed to appeal to the Reichstag and state officials. In the final point of their resolution, they called openly for state action: "In so far as the businessmen are determined to do what their obligation as citizens requires, so they expect from the organs of state government and the appointed representatives of the nation the necessary support, by means of the aggressive implementation of existing laws as well as by means of changes to existing laws, which have permitted the decline of the sense of justice and duty among the working population and the proclivity toward agitation that is hostile toward the state and culture."[75] The concerted response of industrialists in heavy industry therefore served as the model for Bismarck's Anti-Socialist Law of October 1878, which enacted parallel measures of direct repression against Social Democrats. During the legislative debates that produced the law, the decisions of the Saar committee were addressed directly, and employers like Stumm took a leading role in arguing for the necessity of the legislation and its subsequent renewal.[76]

This pattern of local-national repression was paralleled by the close connections between paternalist social policy in German heavy industry and the development of state welfare programs in the 1880s. It was especially industrialists from the large-scale, capital goods, and extractive industries centered in the Ruhr, the Saar, and Upper Silesia, but also employers from the cotton, textile, paper, leather, glass and machine-making industries concentrated especially in southern Germany, who provided much of the impetus behind and support for the interrelated sickness, accident, and old age and disability insurance bills passed by the Reichstag in 1883, 1884, and 1889.[77] These industrialists shared a model of business organization and faced similar economic and labor market conditions: they represented mostly large firms oriented toward the domestic market, supported the tariffs of 1879 as a form of protection for domestic industry and "national labor," offered the most comprehensive schemes of company social provision, and operated enterprises with the highest rates of industrial accidents among workers.[78] They had long favored social welfare schemes, best exemplified by the sickness and pension funds common throughout heavy industry in the 1860s and 1870s, which were organized by firm and governed by a directorate comprising employees under the control of the employer. They opposed both the free funds, organized by occupational status and run by employees themselves, and public or municipal funds, which were also removed from the direct influence of employers, because these schemes offered industrial workers an independent means of self-help and organization and thus weakened the alleged harmony between employer and employee. By the late 1870s, industrialists began to call for national welfare programs in the face of the growing legal conflicts and increasing costs associated with work-related accidents and disability and the independence of the free funds. They also sought to curb Social Democratic influence over the free funds.[79] In these ways, German industrialists combined their efforts to escape the burdens of full financial responsibility and legal liability for workplace accidents and injuries and to reduce the influence of Social Democracy and the trade unions.

The most influential model for the new programs of the German welfare state came, therefore, not from Bismarck and state officials but from industrialists from the Saar and Ruhr in particular, including Stumm, Krupp, Wilhelm Kardorff, and Louis Baare, who promoted their paternalist model of company-controlled *Sozialpolitik* during Reichstag debates over welfare legislation, which they hoped would represent an "extension of proven company programs to the level of the Reich."[80]

When the Anti-Socialist Law was brought before the Reichstag in 1878, these leading voices of German heavy industry warned that coercive measures were simply not enough to contain the growth of socialism; they insisted that more comprehensive and state-directed programs of "positive" social policy were also necessary in order to preempt socialist claims that the state offered nothing but the suppression of workers' organizations. In September 1878, Stumm was the first to propose a motion in the Reichstag, supported by many of his colleagues in the Reichs- und freikonservative Partei (Imperial and Free Conservative Party, hereafter Free Conservative Party), calling on the government to introduce compulsory old age and disability pensions in conjunction with the antisocialist law. He presented these ideas to Vice Chancellor Otto zu Stollberg-Wernigerode and found support for compulsory funds from Staatsminister Karl von Hofmann, president of the Reich Chancellory. The debate over compulsory funds was subsequently taken up by organized interest groups, including leaders of the Ruhr Langnamverein, who invited Stumm to speak at their meeting in January 1879 and subsequently endorsed his proposal. Stumm reintroduced his motion in the Reichstag in February 1879, when his proposal for obligatory funds was formally debated; he was then appointed to the Reichstag commission responsible for drafting a bill for social insurance. The main supporters of Stumm's obligatory funds were drawn from the ranks of heavy industry, including the VDESI, the Langnamverein, and the Bergbau-Verein (Mining Association) from the Ruhr, as well as the CVDI.[81]

Stumm and other industrialists from domestic heavy industry supported a version of state social insurance that differed from the versions of other industry leaders and several leading state officials, including Bismarck himself, who favored compulsory funds run by a national agency and supported with state revenues. This industrialist class fraction favored the social insurance model of the Prussian mining industry and their own large-scale concerns: compulsory funds that were administered by individual firms, were financed by contributions from workers and employers, and distributed benefits according to the level of workers' contributions.[82] They were also concerned to incorporate workers into the machinery of a social insurance program, under the control of employers, in order to take advantage of its "socially integrative" functions. Compulsory funds, as Stumm once explained, would foster the "growing feeling of togetherness between capital and labor, between workers and employers."[83] In addition, these industrialists also sought to end the costly court battles between employees and employers over liability cases involving accidents at work—a key motivation in view of the liability law of 1871, which made employers liable for workplace accidents but required injured workers and their families to prove employer negligence and seek recompense through the courts. This was what prompted Prussian Landtag delegate and Ruhr industrialist Louis Baare to write a memorandum outlining a bill for "no-fault" accident insurance in 1881.[84]

These views corresponded in many ways with the subsequent design of the new social insurance programs, which to a large extent followed the imperatives of representatives from the ranks of domestic heavy industry. The sickness insurance law of 1883, which provided support to workers in cases of temporary illness, did not present many problems for German industrialists, because it was generally conceived as a financial supplement to the more important accident insurance law and left the *Krankenkassen* [sickness insurance funds] of the large-scale factories in place. As Maria Breger points out, employers from domestic heavy industry, under the direction of the CVDI, largely supported the legislation but complained about certain of its provisions. They objected to what they deemed its overly generous offer of benefits, but they were mainly concerned with its failure to eliminate the existing free funds—those independent insurance organizations that were not tied to a particular factory, allegedly enabled workers to fake sickness, and allowed Social Democrats to join.[85] The Reichstag debate over the accident insurance bill, which proposed a system of state-guaranteed benefits for industrial workers injured as a result of work-related accidents, registered a range of interests, emanating from the political parties, social reform organizations, state bureaucrats, and employers from all segments of German industry. But as Hans-Peter Ullmann notes, representatives from domestic heavy industry secured most of their principal aims with the bill, which passed in 1884. The aims accomplished included elimination of employer liability; the sharing of risk through state involvement; employer-employee copayments; employer self-administration through "occupational associations" (*Berufsgenossenschaften*); and, most crucially, the rejection of the proposed "works councils," composed of workers and employers with state oversight, which would have offered workers a limited measure of independence in the administration of accident insurance benefits. According to employers from domestic heavy industry, such work councils would introduce hostility and division at the workplace, destroy the "harmony" between employers and workers, and thus serve the interests of social democracy. The

CVDI prevailed on Bismarck and state officials to strike the works councils from the third draft of the bill.[86]

Finally, German heavy industry supported the old age and disability insurance law of 1889, though not as enthusiastically as the accident insurance law. The initial impetus for this kind of bill came from Stumm, who first proposed a similar kind of welfare provision in 1867 and 1878, but most leaders of German industry at the time favored some kind of old age and invalidity insurance in principle. Their main objection to the new bill was organizational: the government initially proposed that the occupational associations, which were created to administer the sickness insurance, take over responsibility for the administration of these pensions. The *Berufsgenossenchaften* now threatened to become not only an organizational competitor to existing employer organizations but an authority capable of instituting differential pension rates—a payment scheme that would impose heavier financial burdens on mining and heavy industry. But industrialists objected above all to the fact that this kind of administration, with its emphasis on standardized criteria, would limit the ability of employers to administer and control the programs themselves. This threat to established factory pension funds was noted by industrialists from the Ruhr and Saar in particular, but it did not prevent their acceptance of new legislation, especially once the government changed the bill by establishing public (but regional) bodies to administer the pensions in partial accordance with the wishes of domestic heavy industry.[87]

These policies ran parallel to new legislation regulating factory inspection. Employers from domestic heavy industry opposed the kind of state oversight, much more developed in the case of Britain, that came with state inspections of factories and workshops, for the obvious reason that they represented "interference" in the "private" sphere of employers. In 1878, the Reichstag passed a new law extending the remit of state factory inspection to all industrial establishments and workshops with more than ten employees. It created a factory inspectorate, composed of officials appointed by individual state governments, who worked in conjunction with local police.[88] Nevertheless, the introduction of the new accident insurance law in 1884 and the creation of the *Berufsgenossenschaften* allowed industrialists to contain this potential threat to their autonomy. In contrast to the systems of independent state inspection in countries like Britain, the 1884 law empowered the *Berufsgenossenschaften* to issue their own safety regulations and to "appoint their own safety inspectors," who were responsible for making sure that employers adhered to the safety regulations.[89] Employers themselves were therefore put in charge of factory inspection in the context of their own trade associations during the 1880s.

Moreover, the regulatory and disciplinary intentions and gendered bourgeois logic of company paternalism prevailed in most of the debates over state social policy and workers' "protection" during the 1870s and 1880s. As George Steinmetz argues, employers from the ranks of heavy industry successfully ensured the formation of a state social welfare system organized in accordance with bourgeois institutions and intentions and a bourgeois logic of normalization, which sought to inculcate the norms of "thrift," "regularity," "self-responsibility," moral self-discipline, and workplace obedience; to induce the practices and attitudes of working-class "respectability"; and to facilitate the gendered organization of bourgeois family life among industrial workers. The social insurance schemes performed these functions in a variety of ways: they offered benefits only to wage earners, a system that both sustained, and encouraged worker participation in, the broader framework of "commodified" labor and elevated wage earners out of the ranks of the "dependent" poor; they paid out benefits in cash and required self-regulating bookkeeping practices in ways that fostered "a disciplined, planful habitus among workers"; they intervened in workers everyday lives, especially via the local health insurance offices, in order to discipline and rationalize working-class behavior; and they confirmed the general privileging of male industrial laborers—and the male breadwinner norm—by failing to recognize the different working careers of men and women and, in the cases of the invalidity and old age pensions until 1911, by excluding married women workers from access to benefits.[90] In the latter sense, state social welfare schemes followed the social policies of heavy industry very closely. The general bourgeois logic included the gendered definition of work as a primarily male activity, but it refused to acquiesce to the more thoroughgoing proposals of the Center Party and social reformers, mooted in 1885 and 1887, for greater restrictions, if not outright bans, on women's work in industry altogether.[91] Stumm, long a supporter of restrictions on night shifts for women and a determined advocate of women's primary role in the family, rejected such general legal prohibitions on women's work. In this case, the gendering of industrial labor took place within the framework of employer tutelage, capitalist competition, and the "free labor contract," which claimed to leave decisions about whom to hire primarily to the discretion of employers while leaving decisions about where to work to the discretion of men and women workers. This perspective was successfully defended by German industrialists in the legislative debates of the 1870s and 1880s.

Employers from domestic heavy industry were able to achieve most of their aims in the course of these early debates over state *Sozialpolitik,* as well as to shape the overall course of early tariff, financial, and corporate legislation, because they constituted the dominant class fraction during the Bismarckian period.[92] They were able to secure their interests in part by virtue of their organizational unity, through the existence of regional and national industrialist organizations, such as the national CVDI, the Ruhr-dominated or Ruhr-based VDESI, the Ruhr-based Langnamverein and Bergbau-Verein, and the new organizations from the Saar: the VWGWISI and the SGVDESI. These organizations were constantly engaged in public and Reichstag debates over the legislation, by means of petition and publications in their press organs and in the newspapers friendly to industrial interests.[93] Perhaps more crucial, the leaders of German heavy industry had direct and informal access to and connections with the officials in the imperial and Prussian state administrations and Bismarck himself. These connections and contacts were facilitated by means of private correspondence between employers and state officials and numerous private meetings, dinners, or soirees hosted by state officials and Bismarck, occasions that Stumm and others exploited to make known their views on the pending social legislation.[94] They were also forged in extraconstitutional institutions and advisory bodies like Bismarck's short-lived Prussian Economic Council (Volkswirtschaftsrat), created in 1880, which was packed with industrialists and advised the chancellor on the new social insurance legislation during the early 1880s.[95] Finally, employer influence was ensured directly by the presence of industrialists and their allies in the two main industrial parties, the Free Conservative Party and the National Liberal Party. In this capacity, employers from domestic heavy industry not only engaged in parliamentary debate over legislation; they also sat on the important commissions charged with drawing up legislation. Though they were opposed by more numerous firms from the ranks of the export-oriented and consumer goods industries and small business, though social insurance legislation was subjected to a long process of parliamentary debate in the Reichstag over the course of the 1880s, and though its official architects were government ministers and bureaucrats (especially Lohmann, working under the direction of Bismarck), German industrialists succeeded in imposing much of their paternalist vision on the development of state social insurance during the 1870s and 1880s. They viewed the emergent complex of state *Sozialpolitik*—in both its disciplinary and "welfare" dimensions—as largely an extension of their own paternalist practices to the level of the Reich.[96]

Nevertheless, these debates and connections revealed a latent tension between employer efforts to maintain their "private" factory workplace and their involvement in and support for expanding institutions of the state. As a historical concentration of social groupings and discursive fields, the "fissiparous unity" of the German state ultimately opened the paternalist factory regime up to less predictable determinations of party-political debate and official influence.[97] This process was already visible in the debates and legislation over the social insurance programs, which registered the interests of domestic heavy industry most consistently in the debates over accident insurance programs but also revealed the lesser influence of other interests, in the Reichstag and in the state administration, when the version of old age and disability pensions favored by heavy industry was not fully realized. By 1889, industrialists were already expressing some concerns about changing state involvement in the paternalist workplace—the process of interposing legislation "between the employer and the worker" and turning "voluntary" social provisions into state-guaranteed benefits. They were also increasingly alarmed at the influence of "outside" parties on such legislation.[98] As we shall see in the next chapter, as long as this wider set of influences, especially the imperial state and the political parties, was kept to a minimum, the paternalist domination of Saar industrialists in the region was secured by means of their control over local industry and the discourses and institutions of bourgeois *Öffentlichkeit.*

CHAPTER 2

The Public Sphere and Notable Politics in the "Kingdom of Stumm"

In an 1891 speech before a gathering of notables celebrating his own twenty-five years of service as mayor, Wilhelm Meyer presented a potted history of his town of Malstatt-Burbach, centering on the development of the Burbach steelworks and the other key factors in the development of the city: namely, the prosperity of the freight car factory of Gebr. Lüttgens, the cement factory of Böcking & Dietzsch, the coking plant of Dupont & Dreyfus, the machine factory of Erhardt & Sehmer, the iron smelting plant of Friedrich Müller, and others. "All of these establishments," Meyer eulogized, "are in well-guided hands, are making steady progress, and constitute the lifeblood of our development."[1] Emphasizing the benefits of local business for all of the town's residents, the mayor's comments predictably converged with the ideological sympathies of the employer-notables present, but they also pointed to an obvious truth: that the foundations of company paternalism were linked to the institutions and prevailing discourses of public political life in the Saar. This chapter explores the ways in which the guiding metaphor of the factory family, extensive disciplinary rules, invasive social provision, and enforced relation of deference in heavy industry were closely intertwined with and partially constituted in the wider field of voluntary associations, civic activities, municipal governance, and party politics in the region. It argues that local industrialists and their allies forged a paternalist hegemonic order through the narrow confines and exclusionary practices of bourgeois *Öffentlichkeit* in the Saar from the 1860s to the 1890s.

At first glance, the Saar region seems an unlikely location for the formation of bourgeois *Öffentlichkeit,* understood as the social space and collective subject of a critically reasoning public. Contemporary critics such as August Bebel, Friedrich Naumann, and Max Weber, as well as subsequent social historians, have pointed to the overwhelming and suffocating influence of local industrialists and mining officials on public life in the region. The "Stumm system" reduced workers to the status of children, controlled their lives away from the workplace, and silenced the voices of local civil servants, confessional leaders, and social reformers who disagreed with paternalist labor policies—coercive practices that apparently left little room for the development of a public domain of rational debate and opinion formation. Yet the Saar, like other rapidly industrializing regions throughout Germany and Europe, did witness the formation of a regional bourgeois public sphere, especially during the second half of the nineteenth century. In a social-institutional sense, the latter was constituted in middle-class voluntary associations, starting with the Saarbrücker Casino and similar smaller casinos in nearby industrial towns and expanding out into a regionwide network of institutions ranging from charitable societies to economic interest organizations to veterans and patriotic associations. It overlapped with the regulated domain of municipal governance, which included city and town councils and a set of activities, from public ceremonial to council debates, that were closely linked to the wider field of bourgeois voluntary associations; and it complexly intersected with and was born by the main regional newspapers and the networks and institutional practices of the local political parties: the Center Party, the National Liberal Party, and the Free Conservative Party. Moreover, bourgeois *Öffentlichkeit* in the Saar, as in other parts of Germany and Europe, emerged as a "medium" for the discussion of political matters separate from the "material sphere of everyday life, the social conditions of production" and reproduction.[2] In this form, it provided a model of permissible discursive interaction based on the polite deliberation of "self-abstracted" individuals, who claimed to shed their particularity in the process of constituting the unitary "public" and public interest and ignored the structural interconnections between social relations and the public sphere.[3]

Rather than a uniform or coherent social-discursive space, therefore, *Öffentlichkeit* in the Saar was a contested ideological construction, forged in a complex network of institutions and discourses that varied across urban and small-town environments. Saar industrialists exercised considerable influence over the local public sphere and largely excluded workers during this period, but they could never completely silence the representation of other economic interests, especially those of small business. In this regard, it is important to note that *Öffentlichkeit* in the Saar rested on important class differences forged, on the one hand, in the sociologically and ideologically more variegated contexts of Saarbrücken and St. Johann—the two established "Saar cities" with long-standing administrative and marketing functions in the region—and, on the other, in the more homogeneous company towns and mining villages dominated by a small number of large industrial concerns. In addition, the Saar public sphere was structured in gendered hierarchies: it was predicated on the familial model of paternalism, which allowed middle-class women into certain of its deliberations but only in subordinate "feminine" roles. Finally, the deepening confessional divide in the region, which evolved into another axis of social inequality between the mostly Protestant notables and a majority of Catholic workers, was realized in the conflict between a Catholic version of paternalism and the paternalism of local industrialists. In view of these complexities, it makes sense to speak of institutionally and ideologically diverse and potentially contradictory formations of paternalist *Öffentlichkeit* brought together under the hegemony of local heavy industry throughout the Saar region from the 1860s to the early 1890s.

Heavy Industry and the Urban Environment

Paternalist relations of public authority rested on the structural interconnections between the development of large industrial concerns in mining, steelmaking, and glass manufacturing, on the one hand, and the associated rise of an expanded urban infrastructure, on the other, as they were forged and consolidated in the city or town council after the 1860s.[4] As the main governing institution of the locality, the municipal council made the most important decisions over the distribution of city resources and the development of infrastructure. In Prussia, the size of city councils varied according to the size of the population of each city or commune (*Gemeinde*). The municipal ordinance of the Rhine Province required a minimum of twelve councillors to represent communities with no more than 2,500 inhabitants, eighteen in those with 2,501 to 10,000, twenty-four in those with 10,001 to 30,000, and thirty in those with over 30,000. Local statutes, however, could fix the maximum number of councillors, and since this was done by the councillors themselves, much was left to local initiative. This proved to be an attractive means for maintaining limits on the ability of less propertied residents to play a role in public affairs in the industrial towns of the Saar. In 1900, for example, when the population of Malstatt-Burbach was approaching thirty-thousand, members of the town council voted to keep their number at twenty-four.[5] By contrast, town councils routinely conferred municipal "citizenship" rights (*Bürgerrechte*) on newly arrived factory directors and managers in order to include them immediately in the council and in the process of municipal decision making and priority setting.[6] Moreover, as elsewhere in Prussia, Saar councillors were elected on the basis of a three-class voting system. The latter defined eligible voters in terms of property, age, residency, and gender: only men who were twenty-five years of age and older, possessed "independent" means, paid local and state taxes, and owned their own household could vote in municipal elections.[7] This system, which divided voters into three classes according to the amount of taxes they paid, was heavily slanted toward the wealthiest residents and, by design, excluded city inhabitants whose tax assessments were below the requisite minimum; it ensured a very narrow local "citizenry" and a highly restricted "public." In 1876, for example, when Malstatt-Burbach was home to 12,393 residents, only 687 citizens were allowed to vote in municipal elections: 48 in the first class, 161 in the second, and 478 in the third.[8] The numbers of municipal voters in Saar towns did not improve much before 1918. For example, in 1910, Neunkirchen had a population of 34,539, among whom only 5,565 were eligible to vote: 337 of these voters came from the first class, 1,348 came from the second, and 3,800 came from the third.[9]

Accordingly, the municipal councils were sites of limited and regulated *Öffentlichkeit,* comprising mostly propertied and educated notables and some petty bourgeois members, with limited public scrutiny. Municipal matters deemed suitable for "public" consumption were reported perfunctorily in local newspapers, though the extent of this kind of public access varied according to the industrial structure and sociology of the municipality or *Gemeinde.* In factory towns like Malstatt-Burbach and Neunkirchen, council meetings were not readily accessible to citizens beyond the class of local notables. Indeed, information about council meetings and decisions was limited to brief press releases published in approved, industry-friendly local newspapers (i.e., the *Malstatt-Burbacher Zeitung* and the *Neunkircher Zeitung*). This was considered a service for which the newspaper was usually granted an annual subvention, and any "unreliable" or "unfounded" reporting of council deliberations could lead, as in the case in Neunkirchen in 1897, to the exclusion of the newspaper reporter from council meetings.[10] In this sense, municipal *Öffentlichkeit* in the industrial areas was generally limited to the deliberations among town elites. These efforts to limit public involvement contrasted with the more accessible councils of Saarbrücken and St. Johann. In 1885, for example, the mayor of St. Johann proved far more receptive to the idea of broader public awareness of city council business. In response to a petition from a local civic association asking that more extensive reports of council meetings be made available to the press, Mayor Falkenhagen confirmed the right of public access to council meetings but pointed out the limitations of space and the standard practice of delivering short reports of council decisions (and not deliberations, motions, speeches, and voting results). He proposed that the association send its own reporter, whom the council would supply with a desk. The association sent former *Saarbrücker Zeitung* editor Herrmann and obtained more comprehensive reports of council business from then on.[11]

Local administration in Saar cities and towns, outside of Saarbrücken and St. Johann, largely remained the preserve of industrialist notables and their representatives and reflected paternalist priorities. In every industrial area, the town council generally included the leading factory owners and directors from the local iron and steel concerns and the glass factories, as well as mining directors and *Steiger* [foremen] from the local collieries. Among the twenty-six councillors of Neunkirchen in 1888, for example, at least fourteen, came from the ranks of local "industrialists." These included Stumm, his factory director, and three other upper-level employees of the Stumm steelworks; the two directors and two *Obersteiger* [mine foremen] from local collieries VII and VIII; two owners of brick factories; two brewers; and one owner of a sawmill.[12] In Sulzbach, the site of a major colliery and the Vopelius glassworks, the town council was dominated by mining directors and managers from the latter firm, including Richard Vopelius himself.[13] Accordingly, the organization and composition of Saar municipal councils were shaped by paternalist assumptions and practices. The distribution of council seats generally followed a regionwide convention: local heavy industry was allocated at least one-third of the total. Chairman Ewald Hilger of the Mining Office openly admitted this in court in 1904, when he noted that the mining office paid 180,000 marks per year in the town of Püttlingen and thus had the largest stake in the affairs of the town: "We pay two-thirds of the taxes and claim two-thirds of the representative body."[14] In Malstatt-Burbach, officials of the Burbach steelworks similarly claimed one-third of the seats (or eight out of twenty-four) on the city council. Wilhelm Köhl, director of the steelworks rolling mill, justified this distribution in terms of the company's stake in the affairs of the city. Indeed, Köhl even argued that the firm was actually underrepresented, since 33 percent of the town's inhabitants (sixty-five hundred of twenty thousand in 1893) made their living from the steel mill and since the firm paid 49 percent of all taxes in Malstatt-Burbach. This was a distinctively paternalist calculus: managers of the steelworks regarded the company's directors, managers, and employees (and other town residents) as a harmonious unity. From this perspective, the role of company representatives on the council was to speak on behalf of their workers and others who allegedly benefited from their largesse.[15]

Local industrialists were therefore able to secure their interests institutionally, by occupying positions in the main apparatuses of municipal government. The latter included agencies and commissions responsible for overseeing city and town budgets and the activities and institutions associated with a wide array of urban utilities and amenities, including public and private construction projects (e.g., municipal buildings, housing, schools, churches, bridges, streets, cemeteries, and toilets); gas, electric, and water facilities; sanitation, police, and fire departments; and savings and sickness insurance funds and poor relief agencies. In other words, the municipal councils were involved in everything from planning large-scale utility projects to setting the salaries of teachers and street cleaners. In 1896, to take one example, officials of the Burbach steelworks sat on all but three of twelve municipal commissions, including the budget, building and housing, street construction, gas and waterworks, slaughterhouse, water traffic facilities, poor relief, sanitation, and police affairs commissions.[16] Perhaps the example that best illustrates the close relationship between the steelworks and city administration in this sense is the case of Mathias Raabe. Chief clerk (*Bureau-Vorsteher*) of the Burbach steelworks, Raabe was also a city council member (who sat on the budget, police, and slaughterhouse commissions), chief of the municipal fire department, director of the local savings and loan institution, and a school board member.[17]

Saar industrialists and their representatives were therefore well placed to set budget priorities and to structure the urban environment and institutions around the imperatives of heavy industry. This involved ensuring that the appropriate facilities and resources were available for the operation of industrial enterprises, that the relative distribution of taxes and expenditures favored industrial development, and that certain municipally based welfare and labor arbitration bodies favored or at least did not impede the interests of local heavy industry. The construction of railroads, canals, and thoroughfares, along with the introduction of water, gas, and electricity, were always negotiated around the interests of the big firms. In terms of taxation, the influence of local industrialists was perhaps revealed most clearly in Stumm's successful efforts to prevent the conferral of "municipal rights" (*Stadtrechte*) or status on the *Gemeinde* of Neunkirchen. The first petition for full self-administration in Neunkirchen was submitted in 1876, when the city's population reached ten thousand, but by 1900, when it was the home to some thirty thousand residents, Neunkirchen was still classified as a *Gemeinde*. The major stumbling block was Stumm. The conferral of municipal status would force a reconsideration of the tax burdens assessed to Ober-Neunkirchen, where most of the town's citizens, including workers of the steelworks, lived, and to Nieder-Neunkirchen, where the factory and employee housing were located. While both *Gemeinden* shared urban facilities and amenities (from water to schools), they were discrete municipal entities in relation to business taxation. In this way, Stumm was able to keep taxes in Nieder-Neunkirchen to a minimal level, since the factory generated large profits and since there were very few community facilities (streets, schools, etc.) requiring municipal funding.[18] Stumm successfully exploited his economic and political position to frustrate the formation of a fully self-administered city in Neunkirchen.

Finally, industrialists worked in paternalist fashion to control or influence municipal welfare institutions, especially those designed for workers. The city and town councils possessed jurisdiction over a number of important local institutions involving labor questions, including housing cooperatives, employment exchanges, and arbitration courts for employers and workers in local small industry. This allowed factory owners and council members in Malstatt-Burbach, who were responsible for formulating and governing labor policy in nonindustrial enterprises, to write the statutes of the local court, choose its chairman and deputy, define its voter registration procedures, and choose the electoral commission responsible for determining the selection of its observers (*Beisitzer*).[19] While these courts did not involve workers from the larger factories—who, in any case, could be controlled by means of factory disciplinary codes and threat of fines and dismissal from employment—local industrialists took a strong interest in the operations of local arbitration bodies designated for artisanal and other forms of nonindustrial labor, because they feared that trade unionists would secure a local presence if elected to the managing boards of the courts. In other words, Stumm and his allies considered the courts potential sites of more generalized labor mobilization that would enter their own workplaces.

In addition to their influence over resources and municipal priorities, local industrialists sought to control workers' behavior and activity in the urban environment by means of coercive sanctions. In most cases, this coincided with the paternalist emphasis on orderly behavior and moral probity, and this ambition to construct "respectable" workers extended well into the domain of public policing on the streets of the city or town. Management at the Burbach steelworks regularly threatened employees with fines or dismissal from work and company-owned housing for various criminal infractions, including public drunkenness, the discharging of firearms during holiday celebrations, or other forms of public "mischief."[20] Indeed, the mayor and police officials of Malstatt-Burbach cooperated with company officials by turning over lists of arrestees employed by the firm; and police and factory officials cooperated in investigations to determine which workers were responsible for various offenses.[21] Workers and their families living in the settlements along Hüttenstrasse and Nicolausstrasse in Malstatt-Burbach were especially targeted, and New Year's Eve celebrations became a regular occasion for such joint efforts aimed at preventing what officials of the steelworks referred to as "dissolute activities and indecent behavior." In 1897, the steelworks placed a large number of factory firefighters at the disposal of the local police to form twelve patrols (comprising one constable and three firemen each) capable of policing the town during the night.[22]

More systematically, local firms sought to control various forms of popular culture that violated the temporal rhythms of "industrial time." This is perhaps best illustrated in the efforts of local industrialists in the city council of Malstatt-Burbach to regulate religious festivals. As early as 1884, officials of the Burbach steelworks called on the city council to combine the two separate *Kirmes* celebrations (in Malstatt and in Burbach) in order to limit the festivities to one day. They complained of the difficulties of keeping the factory running for days after the celebrations; the measure, they maintained, was in the interest of the workers because it would secure their continued earnings. The council, dominated by officials of the steelworks, promptly complied. Mayor Meyer pointed to larger "national-economic reasons" for the decision: "not only the shortfall of income [*Verdienst*]" was a matter of concern; controlling the *Kirmes* was also a matter of the "greater costs" incurred by workers' attendance at the festivals, especially those resulting from the ways in which the latter "undermine the well-being of the population and affect in an unfavorable way the already weak sense of thrift among the working classes."[23] In subsequent years, the city council simply left the decision over which date to hold the citywide *Kirmes* to company officials, who generally proposed 4 and 5 August as dates that "would best suit the interests of our firm [*Betriebsverhältnisse*]."[24] Finally, council members voted in May 1897 to cancel the *Kirmes* altogether in response to factory director Hans Seebohm's request, which claimed that the *Kirmes* was no longer "appropriate to the times."[25] In response to criticism from tavern and restaurant owners, who stood to lose a large amount of business from the cancellation, Seebohm pointed out that "one is not here simply for the benefit of the publicans." He then emphasized the centrality of the Burbach steelworks to the life of the city and argued that since the workers constituted the "drinking public" and since the steelworks paid their wages, the factory was the "provider" for the publicans.[26] This policy was then advanced by industrialists throughout the entire region, when, in 1898, the Saar Chamber of Commerce demanded the limitation of all *Kirmes* celebrations, which took place in all "districts" connected to industry, to a single day during the year.[27]

Perhaps most politically salient were efforts to harness the apparatuses of municipal government in order to frustrate any public political opposition from industrial workers. As Klaus Saul has argued, these efforts were undertaken in a number of ways.[28] Commissions within the councils could deny oppositional groups like the Social Democrats access to major public venues or meeting halls. Mayors and local councils could also resort to building, fire, and street traffic codes and ordinances in order to deny workers access to meeting halls and public spaces for assembly. In the case of the Neunkirchen city council, city officials stepped up their enforcement of such codes when miners began organizing in 1889; they denied publicans permission to grant miners access to their meeting rooms in their taverns after detailed submissions of architectural blueprints, shut down meetings taking place in overfilled halls, and refused to allow "open-air" meetings by invoking the potential for the "endangerment and disruption of the peace and public order."[29] In this context, the town councils provided a convenient meeting place in which local employers and municipal officials could exchange membership lists and other information about workers' organizations.

Yet the influence of employers was not only institutionally secured directly, in their control over city commissions and offices or their coercive interventions in relation to workers; it was also registered less directly, in the paternalist munificence and assumptions behind municipal social provision. This was the case of charitable interventions and various kinds of social welfare, jointly financed by local industrialists and municipalities, which extended out into the community to an extraordinary degree. The Stumm steelworks, for example, contributed regularly to the construction and operation of schools and churches and offered an open-ended annual subvention in the 1890s of ten thousand to twenty thousand marks toward "defraying the expenses of the *Gemeinde.*"[30] Indeed, Stumm's wealth was so extensive that the city of Malstatt-Burbach took a loan from him in the amount of 125,000 marks at 3.5 percent interest in 1896 in order to pay for the costs of some of its public buildings and other facilities and to purchase land for the extension of its public park.[31] But the paternalist orientation of this kind of activity was most obvious in the case of the new workers' housing cooperatives, which were created largely in response to the efforts of industrial workers to form their own unions. In 1890, for example, city officials, local industrialists, and "respected citizens" of Malstatt-Burbach announced their proposal for the creation of a local housing cooperative as a means of immunizing workers against the appeals of labor organizers who sought to foster "discord and dissatisfaction" and to shatter the "long-standing trust between workers and their employers."[32] The aim was to create a limited liability institution that would enable workers to purchase their own homes—that is, a cooperative that would encourage workers to leave the larger working-class housing settlements, which were deemed potential breeding grounds of labor militancy. In paternalist fashion, the mayor of Malstatt-Burbach and others described the housing cooperative as a means of turning proletarians into "small property owners" as the first step toward "furthering their moral and material welfare," "awakening [in them] a sense for thrift and order," encouraging "more healthy conditions in families," cultivating their desire for their own individual improvement, and thus steering the "social demands of workers onto peaceful terrain."[33]

Despite the convergence of industrial interests and urban infrastructure and the immense influence that private employers and mining directors exercised over municipal councils in the region, paternalist control was never total during this period. Conflicts of interests emerged within the ruling bloc and within local towns in the very process of industrial urban expansion. This was the case with the conflict between Stumm and Karl Röchling, whose coal possessions in Loraine, Aachen, and Westphalia and interest in extraregional and foreign trade brought him into local conflict and even to a near duel with Stumm during the mid-1880s. The dispute, centering on control over the Saar Chamber of Commerce and Stumm's opposition to plans to extend the Saar and Mosel canal projects, created a public rift between the two men until shortly before Stumm's death in 1901.[34] Local conflicts also included the aforementioned debate over the Malstatt-Burbach *Kirmes,* during which local small business protested the decision to limit the popular holiday to one day in the interests of large-scale industry. Conversely, some municipal council decisions favored the interests of small business. In Malstatt-Burbach, for example, the council's decision to create an arbitration court for nonindustrial employers and workers in 1893 was taken with the support of small-scale industry but against the objections of the leading industrial firms, the Burbach steelworks and the Gebr. Lüttgens freight car factory, which opposed the court.[35] In addition, during the strike period from 1889 to 1893, the actions of local employers often produced new municipal burdens and conflicts. Throughout the district of Ottweiler, a number of mayors complained about the costs to local municipalities of public support for sacked miners during the 1893 winter strike. Such financial burdens fostered opposition to the Mining Office's aggressive policy of dismissing hundreds of striking miners.[36] Finally, organized challenges to paternalist control over municipal resources and priorities did occur sporadically. In one early case in Malstatt-Burbach from November 1893, "voters" from the second and third classes, led by an accountant and an employee at the telegraph office, held meetings before the upcoming council elections to contest the domination of the Burbach steelworks over city resources and what they considered the unfair distribution of taxes and fees. They planned to put up their own candidates, "independent men" who would redress the "privileging" of the steelworks and the overwhelming preponderance of the "factory party" (*Hüttenpartei*) in the council itself. Matthias Raabe and Mayor Meyer managed to quell this discontent, but the very attempt to organize in this way pointed to latent tensions within the paternalist municipal order of the Saar, particularly during moments of wider political mobilization such as the early 1890s.[37]

Associational Life and the Bourgeois Public Sphere

The discursive foundations of paternalist *Öffentlichkeit* in the Saar after the 1860s were secured well beyond the confines of the municipal councils:

they were also institutionalized in the expanding field of voluntary associations (*Vereine*), including social clubs, civic and philanthropic societies, economic interest and trade associations, veterans and patriotic organizations, and their communicative practices and public activities. These associations were dominated by male elites or members of the local and regional "middle strata"—that is, the leading industrialists and their upper-level managers, mining directors and officials, merchants and bankers, and local government officials—whose most prominent representatives were usually Protestant. Middle-class men participated in a narrow range of activities in associations and local and regional public offices with overlapping memberships; and they brought to these activities a shared set of highly restrictive assumptions about the forms and procedures of public life. Nevertheless, the public sphere of voluntary associations in the Saar was structured in relations of class, gender, and confessional hierarchy and inequality, not in mechanisms of outright exclusion. It included workers under the tutelage of employers and bourgeois women under the supervision of middle-class men and in accordance with bourgeois codes of feminine comportment and masculine reason. It also proved open to Catholics and their organizations, which broadly traversed its terrain.

As in other regions of Germany, the class dimensions of the bourgeois public sphere in the Saar presupposed the rise of middle-class sociability and processes of bourgeois class formation, anchored in expanding webs of personal connections and wider social networks forged in the meetings of social clubs, nationalist associations, local offices (including poor relief boards, school commissions, and church committees), city councils, business organizations, and the Saar Chamber of Commerce.[38] The wider social influence of bourgeois elites began with the local casino, especially the Saarbrücker *Casino-Gesellschaft,* an exclusive social club for roughly 120 of the wealthiest and most prominent *Saarländer,* including all of the leading industrialists in the region. The successor to an earlier club formed in 1796, when Saarbrücken was part of the principality of Nassau-Saarbrücken, the casino was the locus of political discussion and "convivial association and scholarly conversation" for the upper reaches of Saar "society" well into the 1870s, after which it dropped its commitment to overt discussion of political matters.[39] It became the model for bourgeois social clubs throughout the region, but in the smaller industrial towns and communities, the connections between larger industrial concerns and the casinos and between the interests of industrialist notables and the local "public" interest were more intimate. In heavily working-class communities like Neunkirchen, Dudweiler, Altenwald, and Malstatt-Burbach, the upper-level managers of local firms and collieries formed socially exclusive casinos, which allowed members to socialize without, according to Malstatt-Burbach Casino chairman Friedrich Pelzer, being "required to pass their time with wage earners, etc., in the taverns."[40] Thus the *Casino-Gesellschaft* of Malstatt-Burbach, which met in its own building on the grounds of the Burbach steelworks, was primarily composed of officials of the latter firm, mining officials from the local colliery, and the "better segment of the Burbach population";[41] the Dudweiler Casino created in 1853 was chaired by a mining director (*Oberschichtmeister*) and dominated by local mining and government officials living in Dudweiler;[42] and the *Casino-Gesellschaft* of Neunkirchen met in a house owned by the Stumm steelworks, was chaired by Stumm himself, and was dominated by officials of his firm and other "leading" members of the local *Bürgertum.*[43]

From the 1860s to the 1890s, the casino was a crucial nexus linking notable sociability, bourgeois "claims" to "social leadership," and industrial-paternalist interests throughout the region.[44] The Saarbrücker Casino, for example, was the location of celebratory dinners sponsored by the Saar Chamber of Commerce. It hosted exclusive dinners for the local military regiment (Seventh *Lancers*) and elite veteran's organizations (*Kriegervereine*); it was the venue for notable celebrations of patriotic holidays such as the kaiser's birthday as well as lectures by speakers associated with the nationalist pressure groups (e.g., the Colonial Society); and it was the point of reception for dignitaries from Berlin who visited the local factories and took part in paternalist ceremonies celebrating their visits. Similar kinds of activities took place in the smaller, local casinos. In Malstatt-Burbach, officials of the Burbach steelworks and members of the *Casino-Gesellschaft* (who were often the same individuals) celebrated patriotic holidays and the public service of local elites. On one revealing occasion in 1897, Matthias Raabe was honored for his long years of service both to the steelworks and to the city in his various capacities as public servant. On another, in 1891, the casino hosted a meeting of town councillors and local notables for the purposes of forming the previously mentioned housing cooperative, jointly funded by the city and local industrial concerns, which would enable industrial workers to purchase their own homes.[45]

Yet, as this fusion of sociability and public ceremonial suggests, the casinos were only a part of a much wider field of associational activity among industrialist notables, institutionalized in a broad array of social clubs and organizations, which set the dominant tone of public life throughout the region. From the 1860s to the 1880s, the links between general bourgeois sociability and the civic engagement of employers were most immediate in the locality. Here, social clubs and organizations dedicated to conviviality and entertainment, including local citizen's associations, carnival clubs, and choral societies, provided spaces and venues for the forging of social ties and networks among employers and other notables. In the towns and cities, records of these kinds of *Vereine* reveal significant cross-memberships. The industrialist Karl Röchling and his sons, for example, were members of the Saarbrücker Casino, the *Bürger-Verein* of Saarbrücken and St. Johann, and the Saarbrücken *Kriegerverein.*[46] In each locality, this universe of associations tended to overlap substantially with public offices and commissions. This was most obviously the case with the municipal councils, as well as the county councils (*Kreistage*), which were composed of the same notable segment of the population. Not surprisingly, the Röchlings sat on the municipal council of Saarbrücken and in the regional *Kreistag.*[47]

Equally important to the formation of regional bourgeois *Öffentlichkeit* and paternalist ambitions, especially in the years after 1890, were middle-class nationalist associations—the Colonial Society, veterans' organizations, gymnastics associations, and shooting clubs. The regional branch of the German Colonial Association (later Colonial Society) was founded in March 1884, during a meeting in the casino building, in order to promote the cause of German colonial acquisition, overseas business creation, and "economic and cultural [*geistig*]" connections between Germany and other parts of the world. By 1885, the social background of its organizers and 150 members reflected the early unity of the local notability—the comfortable sociability and common undertakings between and among government officials and industrialists and their managers. Its most ardent exponents included Stumm, Vopelius, Hans Seebohm, Bergrat Jordan of the Mining Office, and Ernst Wagner, glass manufacturer and chairman of the two main industrial organizations.[48] In the context of meetings and public lectures, the Colonial Society introduced deliberations on German imperialism, including lectures on German colonies in Africa and China, with all their structuring orientalist and racist assumptions, as routine features of bourgeois sociability and local notable politics, in a way that emphasized the centrality of colonial expansion to the health and well-being of local industry.[49] This activity was directed in explicitly paternalist ways when speakers before audiences of workers at the Röchling and Burbach steelworks delivered lectures entitled "The Significance of Our Colonies for Our Industry and Workers" or "What Kinds of Raw Materials Are Delivered from Our Colonies and How Will We Put Them to Use."[50]

Aside from the confessional associations, Saar employers and bourgeois notables also controlled the few extant workers' associations—namely, company-sponsored musical, theatrical, and conviviality clubs and *Kriegervereine*—in paternalist fashion throughout the 1960s and 1880s. As the institutional embodiment of the unity and harmony of the factory family, social clubs like the music association of the Burbach steelworks, created in 1865, comprised company officials—from the factory director down to section foremen—and workers across the full range of skill levels. Their involvement in company celebrations and local events (e.g., patriotic festivities or official receptions of dignitaries) regularly performed the paternalist rituals of deference and loyalty.[51] Similarly, *Kriegervereine,* which were originally established after the wars of unification in order to memorialize veterans' military service and the victories in the wars of German unification, drew employers and workers into mutual paternalist relations of patriotic celebration in the context of association meetings and public ceremonial, including nationalist holidays like Sedan Day and the kaiser's birthday.[52] By 1891, there were thirty-nine *Kriegervereine* with 8,944 members in the German Veterans' League for the Saar, Blies, and Surrounding District (Deutscher Kriegerbund. Saar-, Blies- und Nahe-Bezirk).[53] All of them were organized according to established paternalist hierarchies. The *Kriegerverein* of Dudweiler, for example, comprised miners, artisans, and lower-level officials and was chaired by a mining surveyor; the *Kriegerverein* of Altenwald, with 287 members by 1909, was led by the factory owner Wagner; the *Schleswig'schen Krieger-Verein* of Saarbrücken, which included numerous steelworkers, was chaired by the military officer Sandkuhl and included Hermann Röchling in its executive committee; and the leadership ranks of the *Kriegerverein* of Sulzbach included all of the Vopelius men—the father Richard and his two sons, Louis and Carl—and other local notables.[54] Indeed, the leadership of the regional league was dominated by local industrialists and their allies, including Karl Karcher, the factory owner from St. Johann, as chairman; Hans von der Osten, the chairman of the Saar Chamber of Commerce, as deputy chairman; and Matthias Raabe of the Burbach steelworks and Louis Vopelius of the Vopelius glassworks in Sulzbach, as regular members of the executive committee.[55] Not surprisingly, the veterans' associations were ideologically imbued with the assumptions of employer paternalism. The league not only adopted an antisocialist clause in January 1891; it also took up this battle against the "inner enemy" and "social hatred" by promoting the "virtues" of workers' social subordination and workplace obedience: "only order can govern the world," league leaders admonished, "and . . . satisfaction and happiness can only be attained through work and thrift."[56]

In addition to consolidating and forging class inequalities, Saar voluntary associations, like the other dominant media of *Öffentlichkeit* in the region, institutionalized assumptions about sexual difference and gender hierarchy. While men controlled nearly all such organizations on the basis of norms associated with the masculine exercise of public reason, women were allowed limited access to some institutions of public life, often in subordinate roles. This was most obviously the case with local social clubs like the casinos, which included women as family members or guests of male club members in a wide array of special activities, celebratory dinners, festivals, and family outings. The activities of the Saarbrücker Casino, for example, were explicitly structured around the gendered distinctions of bourgeois respectability: only men could become full members, women and daughters were excluded from the meeting rooms and offered their own "women's salon," and the outdoor and music pavilions and the ballroom were designed to offer spaces for the stylized rhythms of bourgeois etiquette and courtship ritual. However, bourgeois women also entered public life more directly through the local patriotic women's associations (*Vaterländische Frauenvereine*), which started with one organization in Saarbrücken in 1860 and expanded to 4,722 members in six local branches (St. Johann–Saarbrücken–Malstatt-Burbach, Dudweiler, Friedrichsthal, Völklingen, Louisenthal, and Sulzbach) by 1908.[57] Largely run by the wives of leading notables, the Saar *Frauenvereine* engaged in a range of activities related to charity and philanthropy—that is, in performing tasks related to "social motherhood" and deemed suitable for women. These activities were not incompatible with a male-dominated public sphere, since men often presided over the associations' public meetings, during which female leaders and members usually sat in the audience.[58] Rather than excluding them outright during the decades before the turn of the century, the bourgeois public sphere included women in subordinate roles and in particular institutions and enforced strict bourgeois codes related to female comportment as the necessary precondition for their civic engagement.

Employer paternalism was also materialized in the gendered structuring of Saar *Öffentlichkeit* during this period. The metaphor of the factory family and the status of local industrial concerns as family-owned enterprises corresponded with the gendered role divisions and hierarchies of Saar associational life. In a way that paralleled the preeminence of industrialists in the municipal councils, the leading bourgeois associations, and the executive committees of the political parties, nearly all of the leading patriotic women's associations in the region were run by the wives of paternalist employers as well as state officials and clergymen. In 1905, Frau Schwering, wife of the president of the railroad administration in Saarbrücken, chaired the executive committee of the Saarbrücken *Kreis* [county] organization, which, as indicated in the previous chapter, included the wives of the leading employers from the iron and steel, glass, and mining industries—Frau Röchling of Völklingen, Frau Helene Vopelius of Sulzbach, Frau Böcking of Brebach, Frau Weisdorff of Burbach, Frau Wentzel of Friedrichsthal, Frau Bergrat Hilger of St. Johann, Frau Bergrat Jordan of Saarbrücken, Frau Bergrat Johns of von der Heydt, and Frau Obersteiger Ries of Dudweiler—many of whom directed their local branches.[59] These women were involved in everyday charitable activities and certain factory ceremonies—for example, the distribution of Christmas gifts to children of workers or to boys and girls attending the factory schools—in ways that enacted the paternalist familial relationships between the factory "mother" and the children of working-class families.[60] They also secured donations to the *Frauenvereine* from all of the leading industrial firms in the region, for the purpose of financing public events and everyday activities. The Röchling concerns gave money to the Völklingen branch, the Burbach steelworks and Frau Röchling gave to the St. Johann–Saarbrücken–Mallstatt-Burbach branch, and the mining inspection in Altenwald gave to the branch in Sulzbach.[61] In all of these ways, the wives of Saar employers participated in the tutelary relationship of employer paternalism, and the gendered familial understandings of business ownership and the provision of company welfare found their expression in forms of public munificence and associational life involving middle-class women.

Finally, these class and gendered hierarchies were imbricated with the confessional divide within the region, though confessional relations changed from the 1880s to the 1890s. Many voluntary associations emphasized their religious neutrality and, like the Saarbrücker Casino, were open to members from all religions. Local notables and industrialists in particular avoided overt identification as Protestants, in part because they assumed a common class interest with bourgeois Catholics, but also out of fear of antagonizing their Catholic employees and out of a purported reluctance to introduce religious concerns into politics. Nevertheless, the Kulturkampf of the 1870s contributed to the incipient formation of competing confessional associational milieus, anchored in separate Catholic and Protestant charitable organizations and choral societies as well as associations and institutions dedicated to confessionally exclusive sociability. The dominant coding of Saar associational life, however, became increasingly Protestant once again—and confessional divisions intensified—in response to the growing strength of the Center Party and the threat of social democracy in the 1890s. From January to February 1889, local Protestant elites formed a branch association of the Evangelical League for the Defense of German-Protestant Interests (Evangelischer Bund zur Wahrung der deutsch-protestantische Interessen, hereafter Protestant League) for St. Johann, Saarbrücken, and surrounding areas in January/February 1889. The Protestant League sought to combat the "growing power of Rome" and political Catholicism and the alleged "indifference" and rampant "materialism" of the day and became a central site for an expanding Protestant milieu. In defense of "German-Protestant interests" and "German-Protestant culture," it gradually created a common domain of public commitment for regional bourgeois notables—including local clergymen, National Liberal politicians, and most of the leading industrialists and their families (the Röchling, Vopelius, Hilger, and Stumm families)—and their petit bourgeois confessional allies involved in charitable and other initiatives. Across the region, the commitment involved increasing numbers of general meetings, lectures, patriotic celebrations and outings, and publishing initiatives.[62]

At the same time, Catholic priests and laypersons in the Saar established their own voluntary associations and communicative practices, which traversed the dominant, largely Protestant institutions and discourses of the bourgeois public sphere during these years. These activities were mostly sponsored by the Catholic Church and were explicitly designed to shore up the pastoral relationship between the local priest and his mostly working-class parishioners. This began with the creation of Catholic religious and self-help organizations, especially the St. Barbara brotherhoods and the *Knappenvereine* [provident associations], for miners and industrial workers in the 1860s. The St. Barbara brotherhoods, religious organizations run by the local priest, focused on matters related to religious worship, especially devotional practices and ritual, but also sought to "improve" their members by eliminating "immoral" practices in working-class communities, especially the "misuse of alcohol, concubinage, and acts of violence."[63] The local priest also led the *Knappenvereine,* which provided social support to industrial workers, including financial assistance in cases of sickness or funeral expenses for deceased members. The *Knappenvereine* similarly sought to reform the morals of their members and to cultivate workers' obedience to employers and public authority: they sponsored biblical and patriotic readings and lectures from the local priest or teacher, who emphasized the importance of such virtues as devotion, probity, "submissiveness, patience, modesty [*Genügsamkeit*], and duty [*Pflichterfüllung*]," as well as a general acceptance of earthly suffering in the expectation of future compensation after death.[64] By the 1880s, Catholic associational life extended to most of the towns and communities throughout the region and included a diverse array of chorale and musical societies, artisans and journeymen associations, youth clubs, girls' and young women's associations, and charitable and educational organizations, often with their own meetings halls in local church facilities or the local Catholic journeymen's house.[65] Whereas Catholic associational life was situated in complex relations with and often subordinated to the dominant institutions of local *Öffentlichkeit,* it also served to sustain its own version of paternalist relations of public authority—a version that could easily converge with the dominant structures and modalities of paternalist public life in the Saar.

Notable Politics and Saar Paternalisms

This paternalist order in the "Kingdom of Stumm" was also constructed on the terrain of party politics. Well into the 1890s, political life in the Saar was dominated by the coalition between National Liberals and Free Conservatives. For the purposes of Reichstag elections, the Saar region was divided into three electoral districts: Saarbrücken, Ottweiler-St. Wendel-Meisenheim, and Saarburg-Merzig-Saarlouis. Regional political life was centered on the Saarbrücken and Ottweiler districts—largely, it seems, because Saarburg-Merzig-Saarlouis was the most Catholic, least industrialized, least populated, and thus least influential district in the Prussian Saar.[66] The region was therefore dominated by the National Liberal Party, with its stronghold in Saarbrücken, and the Free Conservative Party, led by Stumm and Vopelius, based in Ottweiler. Indeed, in 1887, the two self-styled "nationally minded" or "middle parties" formed a long-standing "cartel," which involved mutual electoral support, joint meetings, and efforts to sustain the ideological common ground between the two parties. The latter rested in particular on widely shared assumptions about the benefits of heavy industry and the labor policies favored by Saar industrialists—namely, the officials of the Mining Office and the owners and directors of the main iron and steel concerns and glass factories. It was practically anchored in long-standing agreements over candidate selection, which Mining Office chairman Ewald Hilger explained in an electoral meeting in June 1893: the cartel would ensure that one seat was held by a mining official, who represented the interests of the Mining Office, and that the two other seats were held by private industrialists or their allies, who represented the interests of private industry. This unity was forged in opposition to the Center Party (and the specter of social democracy).[67]

The two cartel parties were based in different parts of the region but shared a sociology that largely reproduced paternalist relations. The National Liberal Party was sustained primarily by local notables in the industrial areas of the district of Saarbrücken, including Völklingen, Malstatt-Burbach, Dudweiler, Sulzbach, Friedrichsthal, Püttlingen, and the city of Saarbrücken itself. Much like the composition of the party throughout Germany, its most active members were drawn from the ranks of the Protestant "middle strata," ranging from educated professionals, industrialists, and large-scale merchants to small businessmen, retailers, and artisans.[68] Yet the local and regional occupational composition of the party reveals a distinctive social geography: the local electoral associations of Saarbrücken and St. Johann were far more mixed occupationally than were the associations in the industrial towns—where mining officials, industrialists, and their managers dominated the associations during the 1880s—and the representatives of heavy industry dominated the executive committee of the districtwide association of the National Liberal Party.[69] The Free Conservative Party was based primarily in the district of Ottweiler, especially in Neunkirchen, though its electoral fortunes were far more closely tied to its leading national personalities, Stumm and Richard Vopelius, and local state officials. The electoral committee of the entire district of Ottweiler for 1887—a combined list of the "middle parties"—included a large number of factory owners, merchants, mining officials, mayors, and municipal officials but only one master carpenter, four publicans, and one master bricklayer out of a total of seventy-three members.[70]

This social exclusivity was anchored in the deliberative routines and existing institutions of Saar associational life, the networks of local social clubs, trade organizations, and civic and nationalist associations as well as the municipal councils, which liberals and conservatives dominated. The allegedly "nonpolitical" municipal council of the steel town of Malstatt-Burbach, for example, was dominated by National Liberals, who comprised at least nineteen of its twenty-five members in 1890.[71] Similarly, at least (and probably more than) fifteen of the twenty-six members of the Neunkirchen municipal council in 1888 were also members of the recently established National Liberal and Free Conservative cartel.[72] As Thomas Nipperdey pointed out some time ago, most German liberals and conservatives embraced a style of politics that rejected permanent party organization and the routines of mass politics in favor of loose-knit groups of local notables and intermittent campaign activities.[73] This meant that effective party organizations were largely constituted shortly before elections; candidate selection took place usually in small, secret meetings, often in the local casino; and larger "voter assemblies" were then held to confirm the party leadership's selection and announce the party's platform to general acclamation.[74] This kind of "notable politics" could only permit the polite deliberation of the "reasoning" individual, the universal and "enlightened" public actor capable of eschewing self-interest and social particularity as the precondition for speaking about public matters. It was founded ideologically, therefore, on the notion that only the educated and propertied could become political leaders. In the Saar, it converged easily with the paternalist claim that workers would best be represented by their employers, who understood better the needs of workers and the "general interest."

Indeed, leaders of National Liberal Party framed their activities in terms of an ideological discourse that identified the National Liberal Party as the nonideological representative of a unitary public interest. In a typical campaign manifesto from 3 May 1898, the National Liberal electoral committee of Saarbrücken announced its support for a number of particular legislative initiatives, ranging from workers' social insurance to the regulations of certain artisanal trades. Yet the manifesto was careful to insist that the National Liberal Party was not a "one-sided economic party"; rather, "its duty was to represent all classes active in the economic and public life of the nation and, after considering their specific interests, to orient its focus to the common good." In this way, the party centered its message on a general warning to Saar voters of both particularistic threats and the dangers of political extremism—in the form of social democracy, "Ultramontanism" (i.e., political Catholicism), the Poles, and the Far

Right.[75] Indeed, in National Liberal rhetoric, the label "extreme" often became synonymous with "other," since it was applied liberally to Social Democrats, Center politicians, Conservatives, and leftist liberals or Progressives.[76] In the classic fashion of liberal "apoliticism," by contrast, the National Liberals themselves claimed to be politically "moderate" and ideologically disinterested: they allegedly occupied a neutral space—that is, the center—upholding not a particular set of interests but the (unitary) "public" interest and elevating the "nation" or "fatherland" over party and petty factionalism.[77] It was in this sense that the National Liberal Dr. Jaeger described his organization in 1893 as "a party of the general will, not a representative of a social group (*Stand*), an interest group, or a confession," claiming further that "it considers all interests which are raised among the people, unifies or reconciles all conflicting interests, and subordinates all to the well-being of the whole, in which the individual interests can also be accommodated."[78] This ideological claim obscured competing social interests, especially those of workers, and conflated the "public" or "citizenry" with the local *Bürgertum* and the National Liberals, who explicitly called themselves the "core of the citizenry" in 1885.[79] Local party leaders, therefore, embraced a universal orientation that defined National Liberalism not as a specific ideological formation but as the imprint of the disinterested and reasonable consensus itself.

The universalizing claims of the Saar National Liberals were articulated to paternalist assumptions and categories—an ideological formation best understood as a form of liberal paternalism. During Reichstag elections, the causes of the German nation, the military, foreign policy, the colonies, and "culture" were repeatedly linked to the overall prosperity of the German economy and its necessary foundation in the existing paternalist relationship between workers and employers. This was most evident in the repeated invocations of employer generosity and worker loyalty. In 1903, for example, Mining Office director Ewald Hilger pointed out that since "the weal and the woe of workers depend on the election of Herr Boltz [the National Liberal candidate], the workers are obligated to elect Boltz out of gratitude to their superiors."[80] These appeals linked universalizing claims about the general interest to the paternalist vision of the factory family, which was invoked by officials of the Burbach steelworks in support of the same candidate in 1903.

> Workers! We have been forced to rely on common effort that can only be realized on the basis of mutual trust. Just as in a family there

> is a head who cares for everyone and must consider the well-being of the entire family, so are we, your employers, called upon to consider your welfare by providing wage-earning jobs and in every other way. Our interests are also your interests. In the first instance, it is for us and you a matter of the prosperity and flourishing of our steelworks. Social Democrats and the Center Party in the Reichstag have only special interests in mind, so we must ensure that a man who represents the general welfare and the fatherland above the parties is sent to the Reichstag.[81]

Moreover, the implied threat inscribed in such paternalist expectations was never far below the surface. In 1887, for example, Burbach steelworks director and National Liberal committee chairman Hans Seebohm warned other members of the electoral committee of Malstatt-Burbach of the Center Party's attempts to introduce "divisions" in the electorate: "Above all, make it clear to misguided workers that the patience of employers may one day come to an end, and the latter could turn to measures that departed from their heretofore well-meaning disposition. Who would guarantee the support of the poor families of these misled workers—these unconscionable agitators?"[82] National Liberals saw no contradiction between putatively selfless and impartial claims to represent the "reasonable" general interest, on the one hand, and the paternalist habit of speaking from a position of "natural" authority over workers and issuing threats, on the other.

The Free Conservative Party was more openly "Christian" and more directly coercive in its paternalist orientation than the National Liberal Party. It centered its ideological message on loyalty to the kaiser, praise of the army, national unity, confessional "peace," Christianity, and, above all, economic and social policies designed to protect "national labor." An electoral notice of 1889 invoked the "unity and greatness of the Reich under the guidance of a strong royal hand" and stressed the party's support for legislative "measures that protected national labor and benefited the working classes." It praised Stumm for his extensive factory social provision, which anticipated the welfare schemes of the Bismarckian government, and for his paternalist interventions at the workplace: "he always maintained order and discipline at his factory; he knows how energetically to combat the incitement of diligent workers by Social Democratic agitators [*Verführer*]; he has generally cultivated good, Christian morals; and he has helped to turn churches and schools into nice homes [*schöne Heim-*

stätten]." In these ways, Stumm was portrayed as a staunch defender of the monarchy and the German nation; a paternalist provider of jobs, whose commercial efforts had opened up "fresh sources of sustenance"; a protector of working-class families; and a guarantor of moral-political order more generally.[83]

In a much more open and explicit way, Stumm and the Free Conservatives presumed to speak on behalf of workers and instructed them on specifically how to vote. By contrast with the more indirect notices of National Liberals, exemplified in the notices at the Burbach steelworks that called on workers to act as "honorable men," Stumm posted unsubtle threats to employees on the main entrance of his steelworks in 1887.

> Your patriotic sense must again ensure that nothing prevents you from voting unanimously for the highly respected Reichstag delegate and *Regierungsrat* Bormann, to whom the factory is especially indebted. *Whoever does not do this* . . . is responsible for the inevitable collapse of well-being and income of all of us and deserves *no longer to be an employee of the firm.* I therefore determine that it is a duty to God, kaiser, and fatherland, but also to the factory and yourselves, that *all of you without a single exception* vote for *Regierungsrat* Bormann on 21 February.[84]

He also notified his managers that they were to use all available "means" (*Kräfte*) to put the "appropriate" pressures on the workers in order to secure their votes for the Free Conservative candidate Bormann.[85]

Despite these differences, the two cartel parties cooperated quite closely when it came to controlling the voting behavior of miners and industrial workers. In the wake of the regionwide blacklisting system that targeted Social Democrats in 1877 and the subsequent refusal to tolerate any socialist or trade union activity within Saar heavy industry, local industrialists and officials of the Mining Office effectively simply banned all oppositional political activity among workers, on pain of transfer to less remunerative employment, loss of bonuses, wage reductions, or outright dismissal from employment. Stumm even included an antisocialist clause in his factory work rules, which state that anyone who "has in his possession or distributes Social Democratic materials, participates in Social Democratic meetings or associations, or frequents taverns in which Social Democratic meetings are held or publications are available" would be subject to dismissal.[86] The Mining Office similarly banned all employee participation in meetings of the Sozialdemokratische Partei Deutschlands (hereafter SPD) and trade unions and declared certain taverns—suspected of allowing meetings to be held on their premises or making available oppositional publications, including those of the Center Party—off-limits to miners.[87] In the absence of sealed ballots, which were introduced only in 1903, National Liberals and Free Conservatives were also able to police workers as they went to polling stations, in order to prevent them from voting for the SPD or the Center Party. Local town notables, employers, and their managers staffed balloting stations in the industrial areas of the Saar, including Sulzbach, Altenwald, Malstatt-Burbach, Dudweiler, Neunkirchen, Gersweiler, Püttlingen, Altenkessel, Wiebelskirchen, Hühnerfeld, and Ottenhausen. This meant that mining officials, employers, and their allies could readily observe workers as they voted. So, for example, the two polling stations located in the areas of Hüttenstrasse and Nikolausstrasse in Malstatt-Burbach, where most of the voters were employees of the Burbach steelworks, were chaired by Rudolf Korten, director of the steel mill, and Heinrich Wagner, a company official during the elections of 1903.[88] In this context, National Liberal employers and managers crowded into entryways to the stations, handed party ballots to workers, kept close watch on them to ensure that the correct ballots were dropped in the boxes, and even took notes as their employees entered and left the balloting stations. In subsequent court testimony, one witness complained, "Sometimes the station is so full that there are more observers than voters."[89] Local state officials, who were often members of the two cartel parties, colluded in the suppression of SPD and Center vote counts, routinely exchanging information and lists with local employers, searching the homes of suspected Social Democrats, and contributing to the exclusion of socialists from the region.[90] After the 1898 election, mayor and National Liberal Hermann Offermann of Wiebelskirchen, for example, gave the director of the local mining colliery, a member of the board of direction of the local National Liberal electoral committee, voter lists on which persons who voted for the Center Party were indicated in red.[91] Finally, these coercive measures aimed at banning opposition were accompanied by active efforts to round up workers, from the mining dormitories or individual sections of steel factories, and lead them in "troops" to the polling stations to vote for National Liberal and Free Conservative candidates.[92]

These activities were combined with a certain ideological labor as well. Local employers took advantage of paternalist venues to instruct workers about election issues, political parties, and specific candidates.

Stumm favored the ceremonies for bonus distribution, during which he routinely rewarded workers for choosing the "appropriate" candidates and warned them about the dangers of radical "agitators." During one such ceremony in June 1893, Stumm told his workers that he was pleased to be able to tell the kaiser "personally" about his own victory and the "excellent behavior of Neunkirchen's workers during the election."[93] Officials of the Mining Office were able to instruct miners through their lower-level management, especially the pit foremen, and also through the official press organ of Saar mining, the *Bergmannsfreund* (The Miner's Friend). During the Reichstag election campaigns of 1893, for example, the editor of *Bergmannsfreund,* Ewald Hilger, not only indicated that miners should support the military bill but also openly called on them to support the National Liberal and Free Conservative cartel candidates, Boltz and Stumm.[94] Leading industrialists, including Stumm of course, dismissed claims that they influenced workers in their positions as employers; the opposition, they maintained, were simply unable to admit that the workers were patriotic and loyal to their employers.[95]

Despite this electoral coercion, the Center Party did emerge as formidable opposition, even if it embraced its own version of paternalism. First formed in 1871 in the region, the Center Party was based firmly in the rural electoral district of Saarburg-Merzig-Saarlouis. It later found support in the district of Ottweiler, where it eventually won the Reichstag seat in 1903 and 1912, and more gradually in the Saar cities of Saarbrücken and St. Johann, where Catholic leaders, mostly clergymen and lay professionals, were based. They established their own newspaper, the *Saar-Zeitung,* in 1872 and subsequently began publishing the much more influential *St. Johanner Volkszeitung* in 1884. The principal early architect of the Center Party in the Saar was Georg Dasbach, a populist chaplain from Trier. Prevented from becoming a priest in 1875 as a result of the Prussian Kulturkampf laws, Dasbach became a tireless political activist, establishing a press and publications network and speaking and organizing for the Center in the region around his home base in Trier but also in the Saar. In addition to the *Saar-Zeitung* and the *St. Johanner Volkszeitung,* Dasbach published calendars and pamphlets for working-class readers; and his early success as an organizer for the Mainz Catholic's Association during the Kulturkampf was followed by less successful efforts to create Catholic associations for miners and factory workers in the Saar during the 1870s and 1880s.[96] Dasbach, along with the lawyer Dr. Muth, who established the first Center organization in Saarbrücken, was the key early presence in Center electoral meetings and activities from 1874 to the early 1890s.[97]

The politics of Dasbach and other Center Party leaders in the Saar were premised on paternalist understandings of a pastoral relationship between church and lay leaders and their working-class flock. Whereas the electoral associations and public activities of the National Liberals and Conservatives were dominated by industrialists and other notables, Center organizations were run largely by priests and middle-class professionals from the 1870s to 1914, though with important differences to be discussed shortly. Moreover, the social orientation of the Center Party was inherently paternalistic over this period, particularly as it was defined in relation to social democracy. Center Party leaders warned Saar workers away from militancy and strikes, sought "reconciliation between the individual classes of society," and viewed religious piety as the best means of combating social democracy.[98] The Center position on the "social question" and the "worker question," first addressed in 1876, criticized the "religiously indifferent" free trade policies of the liberals, and local Center Party leaders henceforth became leading advocates for state welfare programs and protective legislation, including the protection of *Handwerk,* as the principal means of alleviating the plight of workers.[99] Its 1881 program announced support for lowering taxes and distributing them more equitably, the "protection of Handwerk against industry," and the "protection of the worker in cases of accident and old age"; and during the elections of 1887, in which battles were fought to a large extent over the military budget, the Center called for more legislation in the interests of the "well-being of the working class."[100] This kind of approach assumed the benevolence and generosity of social superiors and rested on established relations of hierarchy between priest and flock. In this regard, the Center depended on the growing majority of Catholics in the region, particularly among industrial workers, and on their willingness to accept the leadership of their social superiors in the wake of the Kulturkampf. This strategy resulted in the slowly increasing electoral strength of the party in the 1880s.

Nevertheless, Center Party organizations drew on the contributions of industrial workers in more direct ways. Whereas the National Liberal and Free Conservative parties included no workers in their activities except as passive audience members in some election meetings, the Center organizations included shopkeepers, publicans, artisans, and even some workers in their lists. In the 1890s, after the miners' strikes, party leaders began to include industrial workers on some committees. In 1893, for example, the Center's electoral committee backing the candidacy of Graf von der Schulenberg in the district of Ottweiler was chaired by the clergyman Bourgeois and listed 160 members, among whom were thirty-two miners (active, pensioned, and sacked) and at least three industrial workers.[101] Moreover, activists like Dasbach, who sought to attract miners and factory workers in his publications, became so popular by the 1880s that they were routinely sought out by Saar miners with complaints about working conditions. This practice was reflected in Center Party election meetings, which included some participation by workers; and in the increasing crackdown of Saar employers on workers who read Center and Dasbach publications. Thus this early participation of workers, however limited, signaled an important difference from the practices of the National Liberal and Free Conservative parties—a difference that, as we shall see in chapter 4, evolved into Center Party efforts to put forward "worker candidates" after 1903.

Center Party leaders also rearticulated paternalist assumptions within the framework of claims about the violation of the right of workers to associate and assemble politically, and they even invoked the freedom of the individual in their political struggle against the cartel parties. Though this was done primarily to secure the influence of the church hierarchy, priests, and bourgeois laypersons over the worker flock, it eventually helped to dislodge the paternalist authority of local industrialists and their allies within the National Liberal and Conservative parties in the region. Indeed, Center Party leaders filed their first legal challenge to the coercive practices of local industrialists and their allies after the Reichstag election in the district of Ottweiler in 1881. In defense of the right to free political expression, they charged supporters of Stumm's party with various acts of illegal electoral influence, involving the surveillance of miners in polling stations by mining officials and the violation of ballot secrecy by a local state official. The results were subsequently nullified by a Reichstag commission in 1884. This kind of recourse to legal protest proved to be an effective weapon for the Center Party in the Saar during later elections, including the Reichstag elections of 1893, 1898, 1901 (necessitated by Stumm's death), 1903, 1907, and 1912. It not only put pressure on the cartel parties, which were forced to disguise their coercive practices; it also resulted in the invalidation of the results of two elections. Thus the election of the industrialist ally and National Liberal Heinrich Boltz in the district of Saarbrücken in 1898 was invalidated by a Reichstag commission in 1902; and during the same year, the election of Stumm's successor to the Reichstag seat in Ottweiler was similarly rendered invalid.[102] Formal legality, the relative integrity of the German electoral system, and the commitment of most state officials and political parties to the principle of competitive elections, as Stanley Suval and Margaret Anderson argue, meant that oppositional parties like the Center could actually contest local structures of authority.[103] Indeed, as the protests of the Center suggest, Saar paternalism rested not on total control over the local public sphere but on a wider balance of discursive and other social forces—an environment that decisively shifted and opened up possibilities for reform in the context of new state initiatives and worker protest during the early 1890s.

CHAPTER 3

Challenges to Paternalism: The Battle over the Insubordinate Worker from "New Course" Reform to "Stumm Era" Reaction

The first comprehensive challenges to the "Stumm system" in the Saar were linked to the changing relationship between the state, the public sphere, and the industrial workplace during the 1890s. The decisive moment came when Kaiser Wilhelm II announced a new orientation for state *Sozialpolitik.* This "New Course" entailed Bismarck's resignation and the appointment of Leo von Caprivi as chancellor and a fundamental departure from the underlying principles of Bismarckian social policy: it favored regulatory measures that were designed to facilitate labor conciliation rather than paternalist coercion and containment. Introduced to the Reichstag by the new Prussian commerce minister Hans von Berlepsch, the core of the New Course was a set of workplace regulations and institutions mandated by the Industrial Code of 1891, which brought the central state into more direct engagement with the workplace "rights" of industrial workers. This opened up vital political space for Saar miners and workers, who began to contest the conditions of work in the state-run collieries and the privately owned factories from 1889 to 1893. They did so by creating their own organization, the Legal Protection Association for the Miners of the Upper Mining District of Bonn (Rechtsschutzverein für die bergmännische Bevölkerung des Oberbergamtsbezirks Bonn, hereafter Rechtsschutzverein) and new forms of proletarian *Öffentlichkeit.* The latter were anchored in the discursive structures of the bourgeois public sphere and the miners' own "contexts of living"—the ensemble of everyday rhythms and forms of symbolic expression that characterized social interactions among workers in the collieries, factories, and working-class communities. This proletarian "counterpublic" evolved as a "parallel" arena in which miners and workers were able to "formulate oppositional interpretation of their identities, interests, and needs" in ways that ultimately violated the discursive modalities and structuring parameters of the bourgeois public sphere.[1] It proved short-lived but prompted new responses from bourgeois social reformers, with important consequences for politics in the region.

Rejecting the overtly coercive intentions and practices of factory paternalism and against the backdrop of the New Course, a diverse array of Catholic and Protestant clergymen and Center Party and liberal reformers began to experiment with new strategies of social reform that would target the subjectivities and intimate lives of Saar workers in order to invite their active consent to workplace hierarchies and to prepare them for orderly participation in the bourgeois public sphere. Toward this end, the regional Catholic clergy and bourgeois laity created two new organizations, the People's Association for Catholic Germany (Volksverein für das katholische Deutschland, hereafter Volksverein) and the League of Catholic Miners' and Factory Workers' Associations (Verband der katholischen Berg- und Hüttenvereine). Protestant clergymen and local reform liberals responded with the formation of the Protestant workers' associations (*Evangelische Arbeitervereine*) and a new interest in social reform, particularly the ideas of Friedrich Naumann, one of Germany's leading liberal Protestant pastors, emanating from a local reform organization called the Artisans' Association [*Handwerkerverein*] of Saarbrücken. This turn to social reform was conducted largely on the rhetorical terrain of paternalism itself: it was animated by an insistently gendered and moralizing preoccupation with the internal relations of the working-class family and sought to induct Saar workers into the values and ideological commitments of a religiously informed middle-class lifestyle (*Bürgerlichkeit*). But rather than resecuring a hegemonic paternalist order guided by Stumm and his allies, the new reform movements inaugurated a period of intense political conflict over the shape of the industrial workplace within the local *Bürgertum* during the mid-1890s.

These conflicts were part of a wider debate over the direction of *Sozialpolitik* taking place nationally during the so-called Stumm era of the middle to late 1890s in imperial Germany. The resignation of Caprivi and his replacement with Chlodwig zu Hohenlohe-Schillingsfürst as chancellor in October 1894, along with the replacement of Count Botho zu Eulenburg with Ernst von Köller as Prussian interior minister, marked the end of the

Berlepsch reform initiatives and the renewed ascendancy of Stumm and other conservative industrialists. The latter were able to reestablish some of their former influence over the new government officials during the second half of the 1890s and attempted to undo the consequences of New Course reforms by means of new repressive antisocialist and antiunion legislation, including the Revolution Bill of 1894–95, the Prussian Association Law of 1896–97, and the Penitentiary Bill of 1899.[2] But these attempts called forth new and countervailing developments, which challenged this paternalist reaction: the unprecedented proliferation of competing social reform discourses and initiatives outside the paternalist model.[3] This chapter focuses on these challenges from the New Course to the "Stumm era" and explores their destabilizing impact on the main institutions of the paternalist public sphere in the Saar during the 1890s.

The State, the New Course, and Labor Protest

In February 1889, Kaiser Wilhelm II signaled his desire to become a "workers' emperor" and formally broke with the main principles of Bismarckian *Sozialpolitik.* In a speech before the League of German Occupational Associations (Verband deutscher Berufsgenossenschaften), he announced that it was his duty "to make sure that the workers are guaranteed the protection that is coming to them" and that all "workers" must be guaranteed "legal equality" and should "feel themselves to be a fully empowered estate within the state, and as such they will be recognized by all."[4] One year later, the kaiser put forward a series of proposals for regulating the conditions of work in German industrial enterprises, known as the February Decrees, which were then turned into a set of legislative initiatives (whose principal architect was the newly appointed Prussian commerce minister Hans von Berlepsch) and brought before the Reichstag as a new industrial code bill in May 1890.[5] The bill, which drew on previous Reichstag legislation calling for labor protections, introduced numerous workplace reforms: safety rules and requirements for the proper maintenance of industrial establishments, restrictions on the employment of adolescents and children, and restrictions on women's work, including a ban on night shifts, the introduction of a maximum eleven-hour day, shortened Saturday hours, extended lunch breaks during the workweek, and unpaid leave for new mothers and women about to give birth. It also called for new provisions regulating workplace negotiations between employers and employees: industrial courts (*Gewerbegerichte*), formed on the basis of parity between workers and employers in small industry with state oversight in local communities throughout Germany, in order to provide mechanisms for resolving labor-related disputes; and mandatory works councils (*Arbeiterausschüsse*) to facilitate industrial bargaining between workers and employers in large-scale industrial enterprises. In this regard, the bill introduced provisions governing the content and presentation of factory work rules (*Arbeitsordnungen*). In an effort to curb the punitive "arbitrariness" of employers, it established restrictions on the rights of employers to sanction workers for behavior away from the workplace, a standard maximum for workplace fines, and the requirement to post work rules on the premises of factories employing twenty or more workers, in order to ensure that all employees were aware of company regulations and to force the latter to conform to the dictates of legality. Finally, the bill called for an expansion of the state factory inspectorate by increasing the numbers of state safety inspectors, who were independent of the employer-controlled *Berufsgenossenchaften,* and authorized the creation of the Imperial Commission for Labor Statistics, responsible for gathering information on conditions in German industrial enterprises.[6]

In contrast to their role in the passage of the social insurance bills of the 1880s, German industrialists, especially those associated with the CVDI, opposed key components of the new legislation. Their objections to some of the restrictions on women's work focused mainly on the generalizing tendency of the legislation, which applied to all industrial concerns, and were in part satisfied by the decision to allow the *Bundesrat* [Federal Council] to make exceptions in individual cases. Most industrialists were able to accept these restrictions because they were framed in terms of the gender framework of paternalist practices and were long favored by industrialists like Stumm.[7] But they vehemently opposed provisions calling for increasing state oversight and the creation of institutions for worker self-representation at the workplace. Stumm (whose work rules became the subject of much Reichstag debate) and other industrialists and leaders of the conservative parties objected to the proposed regulation of factory work rules, defended the right of employers to control the political and other activities of workers outside of the industrial workplace, and managed to contain the more far-reaching provisions related to limits on maximum workplace fines and the requirement that work rules be posted in all firms (the regulation ultimately only applied to firms with twenty or more employees).[8] Stumm, Krupp, and Hanns Jencke, general director of the

Krupp concerns, used their influence in the Prussian Council of State, of which they were members, to strip the provision for mandatory work councils from the proposed bill.[9] As a result, the legislation only provided for the introduction of works councils in state-run enterprises and recommended their introduction in private heavy industry on a voluntary basis. Nevertheless, industrialists like Stumm faced a new regulatory regime when the bill was passed in altered form by all parties except the SPD in 1891 and became the new industrial code. The legislation gave state officials new responsibilities in regulating and monitoring the industrial workplace in Germany. Moreover, the kaiser's February Decrees provided legitimacy, however inadvertently, for the self-mobilization of workers, especially in the Saar, where the state-run mines were held up as "model institutions" and where works councils were eventually introduced.

Indeed, when the kaiser received a strike delegation of Ruhr miners in Berlin on 14 May 1889, he helped to set in motion a new dynamic of industrial bargaining between extraregional state officials and Saar miners, who met en masse in Bildstock on 15 May, launched their first strike on 21 May, and formed their own organization—the Rechtsschutzverein—on 28 June. Led by the miner Nikolaus Warken, the Rechtsschutzverein immediately turned into a critical forum for the articulation and representation of miners' interests in relation to the New Course and the kaiser's February Decrees, which became the "credo" of the organization.[10] It also became the principal medium of communication between Saar miners and extraregional state officials. From the beginning, its leaders pressured upper-level mining officials for better working conditions, a course of action that ultimately prompted the formation in mid-July of a state investigative commission responsible for reviewing the written demands and claims of the striking Saar miners in the spirit of "impartiality";[11] in December, they secured an audience with Berlepsch (as *Oberpräsident*) and upper-level mining officials in order to negotiate working conditions and the rehiring of sacked Rechtsschutzverein activists; and they began to work through and energize the "works councils," introduced in Saar mining with elections for council members in early March 1890. When the Rechtsschutzverein sponsored a May 1890 meeting of works council members in Völklingen, its leaders not only described its efforts as the fulfillment of the kaiser's wish that the "Saar mines would be model institutions"; they also drafted a new twenty-four-paragraph petition, the "Völklingen decisions," which they sent to the Reichstag.[12] During the first two years of these protests and negotiations, the Rechtsschutzverein reported impressive membership gains: by November 1889, 6,731 miners had joined the new organization; just over one year later, the figure stood at 20,118; and by the summer of 1891, the Rechtsschutzverein had organized 24,270 miners, or 81 percent of all state-employed miners in the region.[13]

This new public activity and orientation combined with rank-and-file radicalism to challenge several pillars of the paternalist regime in Saar mining. At their 15 May meeting in Bildstock, Saar miners called for the introduction of the eight-hour shift, including winding time; wage increases that would allow each adult member of the work team (*Gedinge*) a shift wage of four marks; a reduction of fines and penalties listed in the disciplinary code; the opening of all colliery gates; the elimination of obligatory contributions to the miners' savings institution; and preferential hiring of miners' sons when possible.[14] The Mining Office initially rejected the miners' demands, but a renewal of miners' meetings and the strike of 11,500 miners on 23 May forced them to make concessions—namely, a new set of work rules that set the daily shift at ten hours, promised to keep the colliery gates open during working hours, set a minimum amount for the team wage, and made payments to the miners' savings institution voluntary.[15] Moreover, after the Mining Office began dismissing Rechtsschutzverein leaders in September and October and the miners began to threaten a new strike in December, the miners' delegation that met with Berlepsch on 13 December secured an official promise that the strike leaders sacked in the fall for their actions the previous May would be allowed to return to work and guarantees regarding wage increases and an eight-hour shift, excluding winding time—promises to be written into the work rules.[16] The following spring, the miners ramped up their challenges to the paternalist regime during the May 1890 meeting of works council representatives, who drew up the aforementioned list of workplace demands called the "Völklingen decisions," which called for the eight-hour day, wage increases, standardized wages, abolition of the *Wagennullen* (excessive fines for mining impure coal), and the opening of the colliery gates.[17] These "decisions" were followed one year later by a new set of demands, the "Altenwald demands," announced by Rechtsschutzverein leaders and works council representatives on 30 April 1890 in advance of another Saar miners' strike from 21 to 25 May. These included the previous "Völklingen decisions" and new demands that the Mining Office rehire all sacked Rechtsschutzverein activists and cease all shipments of coal to the Ruhr region, where miners were on strike. This combination of public organization and militancy from 1889 to 1891 resulted in significant wage gains and permanent limitations on paternalist prerogatives: the annual wages of Saar miners jumped from 976 marks in 1889 to 1,180 in 1890 and 1,212 in 1891, and the Mining Office never rescinded it decisions related to shift times, colliery gates, and sickness funds.[18]

Workers in Saar private heavy industry were also able, if only very briefly, to establish their own forms of independent representation. Like the miners, local industrial workers seized on the kaiser's February Decrees and the New Course legislation and began forming their own "legal protection" associations. These included smaller firm-based organizations at the Wendel coking plant in Hirschfeld in late March 1890 and at the Röchling coking plant in Altenwald in mid-May 1890. But by far the most important new organization was the General Workers' Legal Protection Association (Allgemeine Arbeiter-Rechtsschutzverein, hereafter ARV), formally constituted on 1 June 1890. Led by the local brewery agent Michel Roll, ARV meetings invoked the new worker-friendly orientation of the kaiser as authorization for their deliberations and in response to the efforts of local industrialists to shut down their meetings and force its members to leave the organization; and like the Rechtsschutzverein, the ARV evolved into a critical forum for the discussion of New Course policies and for the expression of workers' grievances.[19] By the end of July 1890, with 224 members, mainly from the industrial areas of St. Johann and Malstatt-Burbach (especially employees of the Burbach steelworks; the machine factory of Dingler, Karcher & Cie; the railroad yards in St. Johann; and the local artisanal trades), the ARV held meetings in which members began to strike at the foundations of factory paternalism throughout the region:[20] they criticized everything from low wages, incentive-based wage systems, managerial arbitrariness in the large factories, and the excessive length of the workday to the disciplinary mechanisms of welfare provisions and the managerial coercion exercised by factory physicians at the Burbach and Röchling steelworks, who allegedly sent injured or sick workers immediately back to the shop floor. They even contested the marriage and secondary employment clauses in the work rules at Stumm's steelworks in Neunkirchen—the most notorious symbol of Saar paternalism. This opposition coalesced in July, when the ARV held two meetings in Neunkirchen, during which Roll and employees of the steelworks launched into a critique of the work rules, working conditions, low wages, the high cost of living in Neunkirchen, and Stumm's bribes and threats to local publicans renting halls to the ARV. Incensed at these attempts to destroy the organization, Roll angrily accused Stumm of buying political influence: "If Freiherr von Stumm did not have such a large money bag, then he probably would not be so fresh. Freiherr von Stumm is no patriot but a briber and a power-hungry man."[21] Nevertheless, the ARV never managed to break free of the social-geographical limitations of the paternalist public sphere. Largely excluded from the company towns where the iron and steel and glass factories were located, it was confined, for the most part, to the more urbanized spaces of St. Johann.

Indeed, without direct connections to and protection from central state authorities, the ARV was easily crushed by Saar industrialists and their allies. Saar employers, who were not subject to legal constraints faced by officials of the Mining Office, were able to destroy the ARV by exploiting the asymmetries of power that characterized the wage relationship and the paternalist structures of the bourgeois public sphere. In May, the executive committees of the VWGWISI and the SGVDESI announced their intention to refuse to employ any "worker who joins trade associations, trade unions, or 'legal protection' associations that are one-sided and aimed at opposing the employers"—a policy that would be enforced by means of blacklists.[22] According to a notice from officials at the Burbach steelworks, membership in the ARV amounted to "direct resistance against the management of the steelworks": "It will, therefore, be the concern of the firm to remove such elements from employment with the Burbach steelworks in the interests of discipline, peace, and order. And it will be left to the founders of this so-called legal protection association to feed themselves and their families in the future."[23] As employers began sacking workers involved with the ARV in May and June, they understood these dismissals as a coercion-free exercise: according to Hans Seebohm, section managers of the Burbach steelworks were instructed to inform their workers, "in a well-meaning fashion, about the reprehensible and ulterior aims of the association and its leaders," but in doing so, they were "to avoid any kind of pressure." "Much to our pleasure," said Seebohm, "this advice has met with success, and the large majority of the workers in question have come around to a reasonable point of view and . . . have declared themselves voluntarily no longer members of the association." Factory employees who "persisted in their blindness" and refused "voluntarily" to leave the ARV, Seebohm warned, "would be immediately dismissed" after the "legally-mandated fourteen-day wage payment."[24] In addition, Saar employers took advantage of their control over the institutions and deliberative routines of local public life in order to crush the movement. Stumm not only sacked employees involved in the ARV; he also pressured local publicans who made their meeting halls available to the ARV, leaned on the Neunkirchen municipal zoning commission in order to get its members to deny the ARV access to certain public halls, and sued Roll for "slander" in order to silence him—a strategy that successfully frustrated subsequent attempts to hold ARV meetings in Neunkirchen after July.[25] These actions relied on the collusion of local officials, including the Landrat in Saarbrücken, local mayors, and the police, who effectively suppressed the speech of ARV members by invoking the Prussian law of association, routinely harassed its members and leaders, pressured local publicans to deny their meetings rooms to the ARV, and gave local industrialists membership lists and other information about association members.[26] By the fall of 1890, Saar employers and their local allies had all but destroyed the organization; the widespread defection of members over the winter of 1890–91 had taken its toll by the time Roll chaired the ARV's last meeting in Ensdorf in early May 1891.[27]

Saar Miners and Proletarian *Öffentlichkeit*

Saar miners were able to sustain their movement until 1893 by fashioning the institutions and deliberative practices of what might be called a proletarian public sphere. They did this in several ways. First, they created parallel institutions of discursive interaction alongside existing bourgeois institutions. In response to local efforts to suppress their meetings, Rechtsschutzverein leaders began construction of their own building in Bildstock in February 1891. In April 1891, they also began to publish their own press organ, independent of the official *Bergmannsfreund* of the Saar Mining Office, under the title *Schlegel und Eisen.*[28] Most striking, during the fall and winter of 1889, Rechtsschutzverein activists formed their own electoral committees in the region and subsequently financed their own (unsuccessful) candidates for the February 1890 Reichstag elections: Warken ran in the district of Saarbrücken, and miner and Rechtsschutzverein deputy chairman Matthias Bachmann ran in Ottweiler and Saarlouis.[29] Second, Saar miners integrated the deliberative norms and institutions of the bourgeois-liberal public sphere with alternative popular institutions, political expression, and participatory forms of democratic practice. Rechtsschutzverein meetings were organized and scheduled by local trustees, but the process of debate and decision making in the organization involved the steady input of the rank-and-file membership. In addition, during the summer of 1890, the miners began to form their own drinking clubs (*Casinos*) or cooperatives, which dispensed beer and schnapps, as a way of creating public spaces that were not subject to the Prussian association law and thus escaped police surveillance. Combining existing forms of working-class sociability with political aims, most of the over 160 miners' casinos that were created by 1893 were led by local trustees of the Rechtsschutzverein and included many strikers who had been dismissed by the Mining Office.[30] Third, Saar miners evolved forms of self-assertion that combined organization with direct action and "rough" and "physical" speech. As a means of preventing employers from preparing for a strike and of building solidarity, they rejected the carefully planned and calculated strike and the legal requirement that they inform their employer, fourteen days in advance, of their intention to strike. "The breakout of the strike," Warken explained, "does not threaten to destroy the organization; rather the strike is the fount [*Schöpfer*] of organization. A strike, if it is going to be effective, must break in like a thief in the night."[31] As mining officials began to sack Rechtsschutzverein leaders and members, employ strikebreakers during strikes, and call on local gendarmes and military units to police the mines, Saar miners turned to intimidation by means of name-calling, insult, and physical violence as forms of collective self-defense. From 1889 to 1893, miners and their family members were often arrested for calling mining foremen "rogues" (*Spitzbuben*), factory owners "sycophants," and strikebreakers "pit rags"; and the leaders of the Rechtsschutzverein and ARV were repeatedly arrested—Warken was convicted on a total of thirty-four separate charges and spent thirty-two months in prison during this period—for "incitement" to violence or "class hatred," "disturbing the peace," "provocative speeches," "disrespectful expressions," slander, and lèse-majesté, often at the mere mention of the kaiser's name during meetings.[32] Not surprisingly, direct physical confrontations between miners and strikebreakers began to take place as early as the December strike of 1889, when strikers at the Heinitz, Dechen, and Gerhard collieries intervened to prevent the distribution of lamps, the roll call, and other miners from descending into the pits.[33]

Fourth, over time, the miners began to formulate their interests and public demands in distinctively collective and class terms. This was visible in the regionwide attempts to establish connections with and to assist other miners in the Forbach district of neighboring Lorraine and the Bavarian Palatinate from November 1889 to May 1890, workers at the privately owned coking plants in Hirchsbach and Altenwald from March to

May 1890, and the workers in other Saar heavy industries during the summer of 1890. It was perhaps even more visible in the connections they established with Ruhr miners beginning in September 1889, when a delegation of Saar miners attended the first national German miners' assembly in Halle; in November, when Warken and others went to the Ruhr to coordinate strategy with striking miners to the north; and in February 1891, when they sent a similar group to the meeting of Ruhr miners' delegates in Bochum, where they met with the leaders of the socialist Alter Verband and the Christian organization "Glückauf" and agreed on a common set of workplace demands. In April 1891, Rechtsschutzverein leaders began making international contacts by sending a delegation to the international miners' congress in Paris, which included representatives from miners' organizations in France, Belgium, Austria, and Britain and focused on developing common strategies—strategies that prompted Saar miners to demand that the Mining Office cease all shipments of coal to the striking regions of the Ruhr in May 1891.[34] In this context, as Horst Steffens argues, Saar miners began to rearticulate their early and "respectful" appeals to established authorities into a class language criticizing "coal barons," "capitalists," and the greedy "Bourgeoisie."[35] Their growing recognition of themselves as miners in opposition to mining officials and their attempts to build solidarity with other miners and workers in the region were revealed as well in the everyday language and symbolism of the Saar miners' movement, in local May Day celebrations, and in the growing interest in social democracy among some of the members of the Rechtsschutzverein, including the editor of *Schlegel und Eisen.*[36]

Finally, Saar miners contravened the gendered divisions and strictures of the bourgeois public sphere—a feature of proletarian *Öffentlichkeit* dramatized by the massive winter strike of 1893. In response to dramatic wage reductions and the introduction of a new set of work rules that failed to meet the Rechtsschutzverein's earlier demands, several thousand miners met in Bildstock in early December and voted to strike on 1 January, an action that began with formal notification and a petition to commerce minister Hans von Berlepsch and the kaiser but quickly evolved into a wider struggle that drew on solidarities forged in Saar mining communities, direct collective actions, and embodied or "physical speech" (*Körpersprache*), in which the body, physical gestures, and violence became a direct means of interest representation.[37] The winter strike was bitterly contested in the pits, in the streets, at the railroad stations, on the paths leading to the collieries, and in fields and open clearings near the mines. Striking miners came into direct and violent conflict with the local gendarmes and blackleg labor: attempts by police to close down meetings, break up pickets, and arrest leaders of the Rechtsschutzverein led to armed crowd actions aimed at securing the release of fellow strikers. In response to official intervention and coercion, the miners broke windows in the homes of "offending" mining officials and those miners who refused to leave the pits; threatening posters were tacked to local buildings and trees, dynamite charges were set before the house doors of opponents, and the revolver became a weapon of intimidation.[38] Most striking was the involvement of entire families, especially women, in the conflict. While a sexual division of tasks was discernible in most of the strike activities of men and women, the miners' wives were nonetheless engaged in distinctly "unfeminine" behavior: they and their children were most responsible for ostracizing and "shaming" potential strikebreakers through insult and intimidation. Women also spat on opponents and were identified with the growing violence; they marched in processions, apparently carried their own revolvers, and broke the windows of strikebreakers' homes.[39] Moreover, during this particular strike, women became "public" in new ways. As early as 1 January 1893, miners' wives were participating in strike meetings, encouraging their husbands to continue the conflict. On 2 January, however, the women themselves organized and conducted a meeting for "wives of the striking miners," the first of several meetings during the first week of January. On 5 January, the high point of women's public mobilization, some ten thousand protestors converged on a packed assembly hall in Bildstock. According to local papers, while their husbands mingled in the streets outside, women spoke about the importance of the strike and solidarity "until they were hoarse."[40] The general radicalization of the miners' wives prompted extreme reactions from local mining officials and bourgeois liberals, who denounced the women protestors as the "purest of shrews," whose "nasty and fanatical manner" exceeded that of the men in the "savagery of their views and speech."[41]

In view of the fact that the miners and their family members had decisively abandoned the path of gradualist proceduralism characteristic of the bourgeois public sphere and had subverted its constitutive separation from the spheres of production and reproduction, it is not surprising that mining officials in Saarbrücken and Bonn and regional and local state officials—with the strong support of local private industry—agreed to break the strike and the Rechtsschutzverein with mass dismissals on 9 January. Accordingly, they sacked some five hundred "principal agitators" and laid off an additional two to three thousand miners. In part because such measures technically did not ban the miners from joining the Rechtsschutzverein and therefore restrict their legal right to organize, long-standing opponents of the miners' organization were also joined by many who were at least partially sympathetic to the miners' cause in the past: most notably, Berlepsch and the reform parties in the Reichstag (the left-liberal Progressives and the Center). Only the SPD defended the miners. In the wake of dismissals and arrests, the strike unraveled, and the miners' movement was weakened by a massive wave of defections by late January. A new cohort of less militant leaders then voted down the leadership of the Rechtsschutzverein, especially those who were receptive to the SPD, and assumed control over the organization. When Peter Schillo, a Catholic hostile to the SPD and the radical elements in the Rechtsschutzverein, took over for Warken in mid-March 1893, he changed the overall orientation and tactics of the organization. Schillo promised to return to the "original" purposes of the Rechtsschutzverein and strict adherence to the pursuit of the "just demands" of the miners "by means of the presentation of well-founded petitions before the appropriate legal entities"—a return to the studied gradualism of bourgeois *Öffentlichkeit.*[42] The radical reorganization of the miners' organization and its subsequent reliance on petitions, which were sent to the Reichstag and the Prussian Landtag but were never answered, failed to save the Rechtsschutzverein. By the end of 1893, nearly all members had given up their membership.[43]

Catholic Social Reform

In response to the radicalization of the miners' Rechtsschutzverein and its distinctively proletarian forms of self-assertion and subversion of the gender hierarchies of bourgeois *Öffentlichkeit,* Catholic clergymen and politicians like Georg Dasbach turned to a new strategy of social reform for the purpose of immunizing Saar workers against socialist and even independent politics after 1890: the active mobilization of working-class loyalties by means of new organizations, the Catholic Volksverein and the Saar miners' and factory workers' associations. These initiatives were conceived in the context of the New Course legislation of Berlepsch, along with the support offered in the kaiser's February Decrees and Pope Leo XIII's May 1891 encyclical *Rerum Novarum,* which represented an official papal attempt to respond to the deleterious consequences of economic and social transformation by authorizing the formation of "workers' associations" for the purpose of providing religious instruction and social support to workers and fostering a "healthy relationship between workers and employers in regard to rights and duties."[44] But it was mainly Center politicians and Catholic reformers within Germany, rather than officials of the church hierarchy in Rome, who led efforts to create organizations capable of pulling Catholics out of the siege mentality of the Kulturkampf era and playing an active political role in defending the rights of Catholics within the Protestant-dominated state, particularly in relation to workers and the social question.[45]

Center Party leaders (namely, Ludwig Windhorst, Franz Brandts, August Pieper, and Franz Hitze) conceived of the Volksverein as the means by which to "school" Catholics "socially" in order to "re-Christianize" the social or to preserve the central institutions of a "Christian social order"—religion, monarchy, family, and private property—as a "bulwark against social democracy."[46] It would do this, as its leaders openly maintained, by mobilizing support for the Center Party, but its main purpose was to serve as a mass-based "propaganda association for Christian social reform." From this perspective, according to Center politician Karl Trimborn, it was necessary to take "Christian social truths" directly to the "communities and families." "These times stand under the sign of the social question," he argued, and "would be negotiated not only in the parliaments but everywhere—in the factories, workshop, etc."[47] Organizers and activists of the Volksverein therefore embraced a more determined approach to social reform, which advocated more expansive social insurance measures and protective legislation for workers and adopted more systematic and populist methods, including general meetings, lectures, journals, and the mass distribution of brochures and flyers. This commitment resulted in impressive early gains for the new organization: by the end of its first year of activity, the Volksverein could already claim three hundred branch directors, 3,133 trustees, and 108,889 members throughout Germany.[48] In the Saar, Georg Dasbach, along with Center Party leaders Peter Döhmer, Dr. Strauss, and Dr. Muth of St. Johann, the priest Bourgeois from St. Wendel, and the priest Laven from Sulzbach, set up a regional office of the Volksverein in February 1891. But only in August 1892, after Catholic leaders gave up on transforming the Rechtsschutzverein from within, did the Volksverein set down roots in the Saar and begin holding mass meetings, involving local leaders and national figures like Trimborn and Center Party Landtag delegate Ernst Lieber.[49] In November, organizers set up a legal aid office, or *Volksbüro,* which offered workers—and other Catholics—assistance in matters related to civil law, including insurance, pension and *Knappschaft* benefits, workplace safety regulations, and taxes; by the spring of 1894, the Volksverein's membership had risen to roughly 2,200.[50]

Local leaders defined the Volksverein in terms of a pastoral strategy for reestablishing the social and political subordination of Saar miners by denouncing their independent activism and Social Democratic sympathies in a highly gendered discourse about working-class moral endangerment, especially in relation to the working-class family.[51] In the midst of the miners' mobilization, Catholic priests met in March 1890 to discuss recent developments and attributed what they regarded as the general lack of "discipline" and declining respect for "spiritual and worldly authorities" among Saar workers to a crisis of moral degeneration or "brutalization" in local working-class communities, especially the unrestrained sexual license between men and women that allegedly accompanied the rise of popular entertainments (dances, shooting galleries, carousels, etc.) in the region.[52] In August 1890, Dasbach similarly castigated the miners for making demands on their employer, rather than politely asking for improvements of their working conditions; called on the miners to accept their religiously sanctioned subordination to their social superiors; and preached the values of "poverty, obedience, and sexual modesty," as well as "self-denial" in relation to "pleasure" and "earthly goods," as the best means of establishing "peace" between workers and employers.[53] Volksverein leaders, determined to defend the "Christian social order" and "throne and alter, home and hearth," embraced this moral-religious explanation of labor militancy and articulated their formal political demands to injunctions about appropriately gendered role division within the working-class family, especially the roles of working-class women as wives and mothers and the responsibilities of the "Catholic man" in the "Christian family." This explains their sense of urgency as they responded by means of a handbill to articles in *Schlegel und Eisen,* the miners' newspaper, in the spring of 1893: criticizing the attempt to turn Jesus into a socialist, the handbill condemned the articles for their contempt for private property and their "attack on the Christian family," a reference to an article in *Schlegel und Eisen* supporting divorce in cases of failed marriages.[54] In this way, Volksverein leaders explicitly articulated together the electoral concerns of the Center Party, social support initiatives for workers, and gendered moral injunctions to "Catholic man" and woman in relation to working-class families and everyday behavior.

Their efforts were buttressed by the new Catholic workers' organizations established under the rubric of the Saar League of Catholic Miners' and Factory Workers' Associations in 1895. These groups were inspired by discussions among leading Catholic reformers about the need to create occupational organizations, rather than trade unions, for workers within existing Catholic *Vereine,* in order to secure the influence of the Catholic clergy and lay reformers over workers directly. They were conceived in the context of a wide range of reform initiatives and drew on existing institutional connections and affiliations, including the Volksverein and the Center Party, since the purpose of the league was not just to create new workers' organizations but to link existing Catholic associations with mostly working-class memberships under a single umbrella organization, as part of a wider attempt at organizational concentration throughout the region. Accordingly, in December 1894, Catholic priests and political leaders, in their capacity as "presiding officers of the Catholic-social organizations of the Saar region," announced their intention to create a new organization, and on 3 February 1895, largely under the direction of Matthias Oesterling, the new league was called into being.[55] By September 1895, its leaders already reported a membership of fifty associations with 7,727 individual members; after slow but steady growth, they claimed sixty-two associations with 9,372 members by 1899.[56] Regular association meetings, communication among its leadership, and an annual league festival served to coordinate its activities and commitments. The latter included modest social programs, including sickness insurance supplements, financial support for funerals, savings accounts, and technical schooling. All of this activity was organized and directed by Catholic clergymen—Oesterling was the president of the league—and bourgeois laity on behalf of workers.

In this way, the miners' and factory workers' associations became venues for local pastoral initiatives, which were designed to disseminate and secure the ideological aims of Catholic social reform more effectively. From the start, leaders of the new organization issued a public statement to reassure the Mining Office and local employers that the league did not constitute a "trade union in disguise," which leaders of the Mining Office publicly accepted. They emphasized its antisocialist intent by barring their members from participation in Social Democratic activities.[57] Rather, their main goals were educative and cultural: they promised to "promote the religious-moral life of, and the sincere harmony among," their members; encourage intellectual development (*Fortbildung*), especially the "correct understanding of matters related to" work or a trade; and represent the "material interests of the members with respect to the peaceful relationships with employers and in consideration of the latter's interests."[58] Toward these ends, the new associations offered a wide range of programs and activities designed to shape the political subjectivities and moral behavior of miners and factory workers, including regular and Sunday meetings, lectures, reading groups, choral and musical clubs, and festivals. Finally, much of this activity addressed the gendered concerns of Catholic social reform directly: the involvement of working-class families in many association events, especially outings and celebrations, and the efforts to disseminate "healthy literature" were all part of the wider aim of "ennobling family life" among the members.[59]

Despite their decidedly paternalist and hierarchical attitudes about the moral deficits and political immaturity of workers, leaders of the Volksverein and the Saar League of Catholic Miners' and Factory Workers' Associations adopted a reform logic that opened up a space of critique. The latter grew out of the populist orientation of the Catholic organizations. First, they appealed to all "estates"—"high and low, clergy and laypersons, workers and employers"—in defense of "Christian society, throne and altar, home and hearth," which necessarily implied compromises from all directions. If workers were to remain obedient, employers were expected to moderate their demands. In the first handbill from 1890, leaders defined the Volksverein as an organization that "wants to make employers and employees more conscious of the obligation inherent in their relationship to each other and pave the way for the recognition of a community of interest among both sides."[60] Second, the emphasis on active engagement with social reform implied some changes to the existing paternalist regime. The new Catholic organizations sought not only "protection against false teaching" but also "the encouragement and active pursuit of the correct principles in the social sphere."[61] This meant the "promotion of social reform, so that the struggle against revolution would take place on religious and social domains" and would be fought not with police measures or "police batons" (*Knüppelei*) but with "practical social reforms"—an orientation that challenged the directly coercive practices of Saar industrialists.[62] In this sense, the Volksverein and the Saar League of Miners' and Workers' Associations could potentially become reformist vehicles for Catholic workers dissatisfied with the labor and social policies of local industrialists.[63]

Indeed, it was in the formal political sphere that Catholic reform initiatives posed the most direct threat to employer paternalism and its defenders in the National Liberal and Free Conservative cartel. The

Volksverein in particular was always understood as a mass-based organization with links to the Center Party. Leading national and local Center politicians like Karl Trimborn, Carl Bachem, Eduard Fuchs, and Franz Schädler were members of and took part in Volksverein meetings, called on audiences to join the new organization and to support the Center Party, and included the Volksverein in their strategic calculations for Landtag and Reichstag elections.[64] In addition, regional connections between leaders of the Volksverein office and the Saar League of Miners' and Factory Workers' Associations were concentrated not only in the figure of Oesterling but also in the wider social networks of Catholic organizations and church offices, especially with the support of Bishop Korum in Trier. This general thickening of local Catholic organizational life became a cause of concern for industrialists and their supporters in the National Liberal and Free Conservative parties, in light of the Center's growing electoral success in the Saar during the 1890s. In this context, the most serious challenge posed by Catholic reform organizations was not their ability to transform paternalist work relations in Saar heavy industry in any direct sense but their wider threat to the electoral dominance of the ruling cartel parties, which could have indirect consequences for local factory regimes. Center Party victories would allow Center politicians to advocate for a range of social welfare and labor protection measures that local industrialists opposed—that is, to secure political leverage in the Landtag and Reichstag, institutions from which incursions into the paternalist industrial workplace were possible.[65] Partly this fear that Catholic reformers could mobilize voters in support of the Center Party drove local Protestant liberals to experiment with their own versions of social reform.

Protestantism and Reform Liberalism

In response to the independent mobilization of the miners' Rechtsschutzverein, the threat of social democracy, and the efforts of Catholic priests and politicians like Dasbach to organize Saar workers, local reform liberals and Christian Social Protestant clergymen also began to experiment with strategies for securing the paternalist aim of a "peaceful relationship between employer and employee." This began in the fall of 1889, primarily among adherents of the Christian Social movement within the ranks of the local Protestant *Bürgertum,* in an effort to collapse the social distance between bourgeois and worker and to generate "more feeling for the people" among Protestant clergymen, liberal teachers and politicians, and local artisans.[66] Drawing on the experiences with the Protestant men and boys' clubs (*Männer- und Jünglings-Vereine*) of the 1860s and 1870s and the model of Protestant workers' associations in the Ruhr, Protestant reform initiatives in the Saar expanded into a wider regional movement of Protestant workers' associations in response to the miners' mobilizations beginning in 1889. Local Protestant pastors and teachers created the first association in October 1889 in Friedrichsthal, not coincidentally the seat of the Rechtsschutzverein, and after one year, they consolidated their six newly created organizations into the League of Protestant Workers' Associations of the Saar (Verband Evangelischer Arbeitervereine an der Saar).[67] In the summer of 1893, the Saar league joined the German League of Protestant Workers' Associations (Gesamtverband evangelischer Arbeitervereine Deutschlands), the nationwide umbrella organization founded in August 1890 by Pastor L. Weber from Mönchengladbach and encompassing six other regional leagues in Rhineland-Westphalia, central Germany, Baden, the Palatinate, Württemberg, and Hesse.[68] By 1900, the Saar league comprised twenty-four associations with forty-one hundred members and enjoyed the support of the Protestant League and the Protestant newspaper in the region, the *Evangelisches Wochenblatt* (Evangelical Weekly).[69]

Aside from opposing the Rechtsschutzverein, the main ideological goal of the Protestant workers' associations was to combat political Catholicism and Catholic reform initiatives and all socialist activity in the region.[70] In 1889, local Protestant clergymen traced the opposition within the Rechtsschutzverein to their own efforts to organize Protestant workers to the influence of "ultramontane agitators," whose efforts to organize the miners allegedly paved the way to "Social Democratic terror."[71] As Dasbach's influence in the miners' organization quickly waned, however, the Rechsschutzverein was consistently associated only with the "tyranny" of social democracy itself.[72] The Reverend Adolf Fauth, a publicist for local political Protestantism and a supporter of the new associations, began publishing brochures about the dangers of social democracy in 1890 and, like other leaders, explicitly defined the confessional organizations as the necessary means by which to promote the "moral-religious regeneration of the entire life of our people" and to "master the terrifying forces of revolution."[73] The main goals of the new associations therefore combined religious, moral, and political concerns: they included the propagation of Protestant beliefs and national loyalties, the cultivation of a "Christian family life," the promotion of "moral elevation and general education," the maintenance of a "peaceful relationship between employer and employee," and the provision of social support in cases of need.[74]

The overall "antiultramontane" and antisocialist mission of the Protestant workers' associations was to come in a putatively nonpolitical program of "education" (*Bildung*) or "enlightenment" (*Aufklärung*)—in the words of the Reverend Nold, a "spiritual and moral, political and social, German-national education"—involving a combination of religious moralizing, patriotic propaganda, and paternalist pedagogy in a wide array of ideological initiatives after 1890. Association leaders offered a richly textured associational life, realized in regular monthly or weekly meetings and periodic regional conferences, political and historical lectures and presentations, religious instruction and Bible study, poetry and literature readings, photographic exhibits, *tableaux vivants,* theatrical and musical performances, youth sections and activities, and family outings and celebrations. The League of Protestant Worker's Associations also established a small library in Ludweiler, in conjunction with the national Association for Christian Popular Education, which disseminated appropriate reading materials in the form of books, novels, magazines, newspapers, calendars, and brochures on a wide range of subjects, ranging from natural science to history. Various kinds of family entertainments, such as the *Festspiele* [plays], with titles like "Luther's Four Days in Frankfurt," "The Kaiser's Dream of Destiny," or "Bismarck as German Defender," reinforced antisocialist and nationalist themes but also provided lessons on a wide array of "moral" concerns related to workers and their families.[75]

This focus on "morals" was most clearly revealed in efforts to promote a "Christian family life," one of the five formal goals of the movement, which rested on explicitly gendered claims about the importance of bourgeois familial order, particularly the moralizing influences of women, as the key to containing working-class radicalism. In his antisocialist pamphlet from 1890, the Protestant reformer Adolf Fauth described Social Democrats as lazy and dangerous "destroyers of the people" (*Volksverderber*), who lured workers into a life centered only on the "enjoyment of all earthly pleasures," including "dances, theater, smoking clubs, and so on," and thus routinely "undermine domestic happiness." Socialist visions of sexual equality and collective child care, Fauth maintained, led to the obliteration of the "Christian family" and the "orderly household." The "good Christian wife" was responsible for preventing this state of affairs.

> Diligently she goes about her work from early morning to the evening, and not a single penny of his [the husband's] wage is spent on unnecessary things; she has nothing to do with finery and knick-knacks; when he comes home from his shift, he finds a simple but tastily prepared meal on the table; it is clean in the living and bed rooms and everything is in good order; the children are obedient and are raised as Christian; they do not cause him any heartache. In this manner, he does not become a tavern brother, but prefers to spend his time at home with his wife and child, where he feels homebound and safe.[76]

By contrast, the working-class woman who lacked domestic skills, indulged in "pleasure-loving" (*vergnügungslustig*) pursuits, and was "obsessed with dressing up" would, according to Protestant workers' association leader Zarth, cancel the beneficial effects of welfare reforms and contribute directly to the political radicalization of her husband.[77] These concerns were taken up in a new series in the *Evangelical Weekly* entitled "Contributions to the Solution of the Social Question" beginning in 1889, during the miners' mobilization; and they were addressed regularly in the activities and programmatic statements of the movement.[78] Association lectures focused on such themes as the virtues of hard work, moderate drinking (and the dangers of alcoholism), and the blessings of a properly ordered family life. Even theatrical presentations, with titles like "Ideal Women Figures of the House of Hohenzollern," included educational content of this kind.[79]

These gendered concerns were not incidental but, rather, central to the political project of Protestant workers' association leaders, because they embraced assumptions about bourgeois acculturation as a precondition for political speech and worked to secure liberal definitions of citizenship and rights, "reasonable" demands for reform, and the boundaries of permissible debate among Saar workers. Association leaders began with the important recognition that workers could no longer be treated as objects of (paternalist) benevolence and tutelage in the age of compulsory education and mass politics. Local reformer and teacher Strunk spoke at the third annual league festival of the Protestant workers' associations about the dangers of social democracy: "The German worker is not the slave who must break his chains . . . the German worker is in every way a citizen endowed with equal rights and equal respect who is entitled to have his wishes and demands considered."[80] This emphasis on legal rights and citizenship was directly linked to proceduralist assumptions about gradual social reform. It was in this sense that Pastor Coerper, another leader of the Protestant workers' movement, argued that the "conditions of our workers need, and are capable of, significant development" and that "here is an area in which the rich fruit of justice [*Gerechtigkeit*]—for example, more secure legal rights, improved economic situation, and a more respected social position and treatment—must be sown and harvested."[81] But association leaders always qualified the exercise of citizenship rights with warnings about more radical political designs. "Nothing really productive," Strunk admonished, "can be gained by way of revolution, however; only on the legal path of well-ordered legislation. Achieving the goals that guarantee citizens satisfaction by way of the legal route is the determined will of the German people and the kaiser."[82] Thus the purpose of the new organizations was not simply to allow workers access to public debate but to reform and channel their political speech and participation into "appropriate" and permissible expressions. Workers, according to Zarth, could be treated as full citizens only after a concerted effort of social reform had offered them the "opportunity to increase their knowledge in an appropriate and purposeful way" and had eliminated dangerous "half-knowledge" or "misunderstood truth." This general program of reform and education would have many benefits for workers, he argued,

> if, for example, they learn here the costs, the terrible misery of revolutions for a people; if they are shown by means of historical conditions that workers today fare much better than in the past; if they hear how economic activity has gradually developed, how in their very existence and prosperity, trade, industry, artisanal production, and agriculture are fundamentally interdependent.[83]

The purpose of the Protestant workers' associations was therefore to cultivate the "private" subjectivities and reform the morals of workers in order to render them as "mature" public citizens, who would make limited political claims and not demand fundamental social changes, as a precondition for their entry into local *Öffentlichkeit.*

Nevertheless, this version of reform contained important, if initially latent, elements of opposition to the paternalist labor regime of the Saar. Whereas local industrialists supported the organizations before 1893, they increasingly grew concerned about the emphasis on citizenship and associational rights and the growing presence of Friedrich Naumann's reform ideas in the associations. Indeed, in the summer of 1893, leaders of the Saar associations were drifting toward the reform group around Naumann and not the conservative faction led by L. Weber in the German League of Protestant Workers' Associations. This became obvious with the publication and local circulation of the Naummanite Reform Program, which addressed conditions specific to the "large-scale factory" and promised to support "workers in their striving for the elevation and improvement of their living situations." The six-point program called for the inclusion of widows and orphans in state social insurance programs, limitations on women's and children's paid employment, the introduction of legally recognized occupational associations or trade unions, the protection of the worker's right to organize, the introduction of workers' representatives and representative bodies in individual factories, and the transformation of state enterprises into "model factories" in a "worker-friendly sense," involving the "guarantee of the full personal freedom of the workers and white collar employees."[84]

This same Protestant-liberal version of social reform was winning adherents among the members of the Artisans' Association in Saarbrücken, many of whom were also involved in the Protestant workers' associations. Originally created in 1864 as an occupational association dedicated to the "intellectual and material well-being of the artisan estate," the Artisans' Association expanded in the late 1880s and early 1890s to encompass a more socially diverse membership.[85] By 1895, the association, led by the gymnasium teacher Dr. Theodor Meyer, had changed its name to Artisans' Association—Association for Popular Education [Handwerker Verein für Volksbildung] and had altered its statutes to reflect its new general mission: the "promotion of intellectual and moral education." Henceforth, the Artisans' Association would sponsor lectures for the general public on "all spheres of knowledge," with the specification that the lectures were to "be generally interesting" and "must not have party-political or confessional points of controversy as their focus."[86] In addition to public lectures, it held regular meetings, built its own library, and sponsored celebrations and family events. This transformation into an association for public lectures and general forum of local conviviality or sociability (*gesellige Unterhaltung*) dramatically increased the size of the Artisans' Association: from 1893 to 1896, its membership rose from about five hundred to eleven hundred.[87]

It also reflected the growing popularity of lectures related to matters of general interest and social reform among the members of the association. Each year, leaders of the Artisans' Association invited prominent academics, artists, and public figures—like the sociologist Max Weber, the poet Emil Rittershaus, and the historian Hermann Oncken—to deliver public lectures, but they also began to invite several leading advocates of social reform, including the economist Adolph Wagner, the social reformer and women's rights advocate Elisabeth Gnauk-Kühne, and Friedrich Naumann. The most publicized talk was Naumann's lecture entitled "Indigence, Charity, and Self-Help Organizations," delivered on 4 October 1894, which called for the comprehensive public organization of poor relief as well as self-help organizations, including trade unions, for all occupational groups, as solutions to the problem of poverty.[88] Less than a year later, Adolph Wagner delivered a lecture entitled "Socialism, Social Democracy, and Positive Social Reform," which addressed some similar themes to an audience of about six hundred people. Rejecting attempts, especially on the part of employer-paternalists like Stumm, to equate the social reform advocacy of some academics with socialism, Wagner insisted that academic reformers proposed many different solutions to the social question, including trade unions of the British kind, self-help cooperatives, protective legislation, and state ownership of essential industries.[89] This advocacy of nonsocialist trade unions was tied explicitly to the gendered imperative to reform the subjectivities and "private lives" of industrial workers as a precondition for granting them the right to participate in the public sphere—a project developed further in two subsequent association lectures, delivered by Naumann and Elisabeth Gnauk-Kühne in October 1895 and November 1895, respectively.[90]

The "War between the Patriots": The Crisis within the Liberal-Conservative Coalition

The new Christian Social orientation within the Protestant-liberal establishment led to intense political conflict between its local supporters and opponents from 1895 to 1898. These disagreements became public in February 1895, when Stumm, Vopelius, and Röchling delivered a warning to the leaders of the Protestant associations about their reform activities, especially the formation of a legal aid bureau in the fall of 1894 and the distribution of Friedrich Naumann's new journal *Die Hilfe* in the associations. Begun as an informal offer of legal advice to workers as a means of reducing the potential of the Rechtsschutzverein-led strike in early 1891, the legal aid bureau evolved into a formal office from August to November 1894, when Protestant Workers' League leaders felt the need to create an organization capable of preventing Protestant workers from using the services of the Catholic Volksverein. The bureau offered workers assistance not only in matters related to sickness, accident, old age and disability insurance, poor relief, taxation, schooling, military service, guardianship, and fire insurance but also in other "matters related to the labor relationship."[91] At the same time, several of the leaders of the movement in the Saar were embracing the ideas of the Naumann wing of the German League of Protestant Workers' Associations and distributing copies of Naumann's brochures to the local associations.[92] This activity coincided with a development even more threatening to local industrialists: Naumann's aforementioned visit to the Artisans' Association in Saarbrücken on 4 October 1894, during which he addressed the topic of independent labor organizations.

Stumm's response to these reformist "intrusions" into the Saar began with his national efforts in the Reichstag to silence Social Democrats by means of exceptional legislation in the winter and spring of 1894–95. During the Reichstag debate over the imperial government's proposed Anti-Revolution Bill, which was designed to curb the "inflammatory speech" of Social Democrats via amendments and supplements to the criminal and military codes and the press laws, Stumm pushed for draconian measures against Social Democrats, including the denial of their "active and passive right to vote" and the punishment of exile or imprisonment of all socialist "agitators." During his 9 January speech, Stumm also criticized academic and religious reformers, attacking what he described as "university socialism" and the dangerous precedent set by Protestant clergymen—that is, the Christian Socials—who "flirted" with the doctrines of social democracy and revolution.[93] Whereas the "socialists of the chair" were gaining a dangerous influence over higher education but at least speaking to "educated audiences," Stumm complained, pastors like Friedrich Naumann agitated among simple workers and spread "dissatisfaction"; they thereby risked turning the Protestant workers' associations into "combative organizations" that would "march directly into the camp of social democracy." Stumm's comments unleashed a storm of criticism from the ranks of social reformers and Christian Socials, especially Adolph Wagner, who called the remarks "irresponsible assertions" and defended the nationalist, monarchist, and capitalist commitments of the so-called socialists of the chair. The controversy was fought out in the pages of *Die Hilfe* and *Die*

Post, and Stumm even went so far as to challenge Wagner to a duel.[94] In the end, however, Stumm did not prevail in the Reichstag, and the last version of the bill, which Stumm himself did not support because it did not go far enough in its restrictions on political speech, went down to defeat in May 1895.

In contrast to this lack of success at the national level, Stumm and his allies, who passed a resolution in support of Stumm during the Reichstag debate, were able to contain, at least in the short term, the activities of bourgeois and Protestant reformers in the Saar.[95] Here, where the immediate occasion for open controversy was Naumann's lecture to the Artisans' Association on 24 October 1895, the matter was more susceptible to the time-tested practices of coercion and inside influence. In addition to silencing local newspapers that printed articles favorable to Naumann by threatening to withdraw his financial support, Stumm pressured upper-level provincial school authorities in Koblenz, via the agency of *Oberpräsident* Nasse, after which Dr. Mirisch, director of the *Oberrealschule,* discussed the matter with Theodor Meyer, who tendered his resignation as chairman of the Artisans' Association in May 1896.[96] Stumm also engineered the removal of two reformist teachers, Dr. Görbig and Dr. Disselnkötter, from their posts at the gymnasium and their transfer out of the region in February 1896.[97] Not satisfied with the extent of the local purge, Stumm, Karl Röchling, Richard Vopelius, and Rudolf Böcking, director of the Hallberg steelworks, attempted to secure the transfer of the director of the gymnasium, Dr. Fischer, whom they suspected of protecting Naumannites within the faculty. They attempted to replace Fischer with Dr. Scheer, a Stumm supporter and teacher at the gymnasium, though the industrialists ultimately failed to prevail in this case.[98]

At the same time, Stumm moved against local leaders of the Saar League of Protestant Workers' Associations. In November 1895, he pressured the reform pastors, who met for their delegates conference on 10 November and both distanced the Saar league from Naumann's reform activities and reaffirmed its commitment to cultivating a "peaceful relationship between workers and employers" and its opposition to social democracy.[99] In view of this ambivalence, Stumm and other Saar industrialists stepped up their attacks on the Protestant clergy in January and February 1896, calling the Christian Social movement a "revolutionary phenomenon dangerous to the public weal" and calling its leaders "conscious or unconscious 'helper's helpers' of social revolution who stand on the same step with Bebel and Liebknecht."[100] After twenty-nine local clergymen issued a public declaration defending their movement in response to these articles, Stumm attacked as dangerous reformers Lentze; Pastor von Scheven, editor of the *Evangelical Weekly,* which Stumm ultimately banned from his factory grounds and welfare facilities in Neunkirchen; and Superintendent Zillessen of St. Arnual. This public accusation prompted the overwhelming majority of Protestant clergymen of the district synod to pass a resolution in favor of Zillessen and the "social activities" of the pastors, as a vital part of their "spiritual office," as well as the publication of a brochure entitled *Freiherr von Stumm-Halberg und die evangelischen Geistlichen im Saargebiet,* written in defense of the Protestant leaders of the associations.[101] Stumm then attempted to pressure officials of the Rhine consistory in Koblenz to initiate disciplinary measures against the offending clergymen and the transfer of Zillessen, Lentze, and von Scheven out of the region.[102] Despite other, similar attempts to use his money and influence—including efforts to have orders conferred on two "loyal" pastors, to pressure the Neunkirchen presbytery into petitioning the Koblenz consistory, to create and finance a new opening for a pastor locally, and to petition the *Oberkirchenrat* [Supreme Consistory] in Berlin in order to demand transfers or disciplinary actions—Stumm was unable to have his way in this case.[103]

These actions ultimately generated a regionwide political conflict over social reform and the worker question, dubbed the "War between the Patriots," which produced a split in the dominant paternalist-liberal consensus and the local "public." The debate between reformers and paternalist industrialists sent reverberations throughout the principal institutions of the local Protestant *Bürgertum.* In April 1896, it entered the "nonpolitical" meeting rooms of the Saarbrücker Casino, the elite social club, when Ewald Hilger was discussing these matters with reform-minded members: "We would have forced out your Meyer, if we had been able to," he reportedly claimed.[104] It also entered the "nonpolitical" halls of the gymnasium, where it divided the proponents of Protestant-liberal reform from the supporters of the industrialists within the local educational establishment. Here the faculty was apparently almost evenly split between a reform faction of eight members, including Director Fischer, who had formed an evening discussion group in order to discuss social issues, and a conservative group, led by Dr. Herwig and Dr. Scheer, comprising nine members who supported Stumm and his allies.[105]

More serious, the *Saarbrücker Zeitung,* the central medium of local *Öffentlichkeit* and the long-standing venue of paternalist interests, became the site of political contestation as local industrialists targeted its coverage of Christian Social views and activities as well as the reform sympathies of its editor Albert Zühlke.[106] Indeed, during the course of the conflict over the Artisans' Association and the Protestant Workers' Associations, the *Saarbrücker Zeitung* had become a leading voice of opposition to Stumm and his allies and even openly challenged the "Stumm system"—"a system," it asserted, "that renders impossible all debate and criticism by means of coercive measures and that wants to suppress all free expression of opinion."[107] In the spring of 1896, the industrialist Karl Karcher and a committee of five self-appointed representatives from the "middle parties"—Boltz, Olzem, Vopelius, *Kommerzienrat* Haldy from St. Johann, and Hilger of the Mining Office—contacted the *Saarbrücker Zeitung*'s owner and publisher, Carl Hofer, to begin negotiations in an unsuccessful attempt to effect a change of editorial direction. The committee offered an annual subvention of ten thousand marks in return for the removal of Zühlke as editor and his replacement with a new editor favorable to the interests of heavy industry.[108] Thwarted, the committee of five proceeded, with Stumm's backing, to arrange for the purchase of the *General Anzeiger,* a local newspaper in St. Johann with no clear party profile and a small readership. Negotiations with a Strasbourg publisher produced an agreement by which the latter would purchase the *General Anzeiger* and receive a subvention of one hundred thousand marks over a period of ten years. The five-member committee would exercise "oversight" authority in order to ensure the new paper's adherence to the political line set by the industrialist wing of the National Liberal and Free Conservative cartel.[109] After a direct financial threat, the previous publisher-editor sold the *General Anzeiger,* and the *Neue Saarbrücker Zeitung,* a newspaper published in the "grand style" and representing the interest of local heavy industry, was officially established.[110] In order to guarantee financial support for the new press organ while at the same time denying support to the *Saarbrücker Zeitung,* Stumm led a campaign to transfer to the new publisher all sources of revenue and advertising from the Saar Chamber of Commerce, the Iron and Steel Occupational Association [*Berufsgenossenschaft*], the companies belonging to the SGVDESI and the VWGWISI, the Mining Office, and local state agencies—the railroad administration, the post office, the tax assessment and forestry offices, the courts, and the city administrations of Saarbrücken and St. Johann.

Despite this success, the actions of Stumm and his allies exposed the workings of power and economic influence in the arena of public debate—

a violation of the authorizing strategies and necessary discursive evasions of bourgeois *Öffentlichkeit*—and thus opened up latent ideological divisions within the local "citizenry" or "public," with far-reaching consequences for the National Liberal Party in the region. When the paternalist leadership of the party negotiated the newspaper deal and called a meeting for 1 March in order to confirm the *Neue Saarbrücker Zeitung* as the official party newspaper and to declare the Christian Social movement incompatible with National Liberalism, it necessarily turned the debate over social reform and the "worker question" into a conflict within the local National Liberal Party itself.[111] The crassness of these kinds of actions, which were routinely employed against workers but generally obscured by the apoliticism of liberal discourse, drew an immediate rebuke from the reformist members of the National Liberal Party. In the meeting, Mirisch, speaking on behalf of the reformist opposition, condemned the attacks on supporters of Naumann by means of "coercion," rather than verbal persuasion, as a violation of the "free expression" of liberal voters. He also criticized the attempt to buy the editors of the *Saarbrücker Zeitung,* for its reduction of political conviction to a salable commodity.[112] A week later, during the first meeting of the reform opposition within the party, the pastor Coerper referred to the *Neue Saarbrücker Zeitung* as the "Stumm organ," implicitly redescribing it as a private publication representing the views of the one man who financed it, rather than a public institution with a wider appeal among the citizenry. This meeting consisted of over seven hundred reform-minded liberal voters, which also included Mirisch, the gymnasium teacher Krämer, the editor Zühlke, and the Reverend Zillessen, who rejected the inauguration of the new newspaper and the actions of the industrialist wing of the party; they ended the meeting with a salute to the "free citizenry."[113] Boltz and the industrialist wing of the party called the next meeting for mid-April in order to seek "clarification and reconciliation" and expel all Christian Social influences from the party. In response, the reformers not only criticized "heavy-handed" paternalist responses to workers' organizations and called for a new approach to the "worker question"; they also rejected the attempts to render reformers from their own ranks "voiceless," Boltz's support for the Anti-Revolution Bill, and his more immediate effort to control the discussion agenda of the meeting. In a moment of political self-reflexivity, the reformers even drew parallels between this treatment and the paternalist domination of workers. "The delegates," Kriene argued, "would not allow the citizenry [*Bürgertum*] to speak in their meeting because they view the latter like they view the proletariat, members of which during election time are addressed as 'dear gentlemen' and 'esteemed assembly' but afterward are held in contempt or, more often, simply not even considered and only expected to remain silent."[114] The efforts of reformers to hold more protest meetings, draw up petitions to higher state authorities, and mobilize their supporters in open rejection of paternalist control over local party politics ultimately threatened to split apart the National Liberal Party organization in the region.

These disagreements were framed in terms of a battle over control of the local citizenry and the "public." The reform faction laid claim to the title of "National Liberal citizen's party," acting, in the words of the shopkeeper Ruhr, in defense of the "independence of the citizens."[115] As general citizens and members of the "citizen's party," as the reformers put it in their 8 March meeting, they raised their "utmost objection to the suppression of free expression, to the politics of coercion."[116] In this way, the reformers counterposed themselves to the "Halberg party" and invoked the general "public" in a way that tried to reclaim the ground of a unified and "free citizenry" demanding the right to "free expression." In addition, they rejected the way in which this "public" was excluded from the decisions made over the creation of a new newspaper; and, outraged by the transfer of the teachers as an unwarranted intervention of the state into affairs of the citizenry, they petitioned state officials and even the kaiser in support of Dr. Meyer. The "independence and trustworthiness of state officials" was explicitly called into question, as were the attempts to restrict the legal rights and political freedoms of the teachers and the broader public.[117] Much in the way it usually served to particularize workers, this rhetorical strategy was now deployed to exclude the paternalist faction of the party from the liberal consensus.

Stumm and his allies in the party leaderships also spoke to and invoked the name of the public. When it appeared that state officials would refuse to allow the Mining Office to transfer the business of the *Bergmannsfreund* to the new newspaper, Stumm wrote to *Oberpräsident* Nasse and claimed that the *Neue Saarbrücker Zeitung* would necessarily appear to be "boycotted in the eyes of the entire public because everyone knows that the promises were made." "Under these conditions," he asserted, "neutrality really amounts to opposition to the 'middle parties' and support for the Christian Socials."[118] Stumm added, "The entire mining population will view this as a change of policy against the new press organ, and the latter must, therefore, appear without backing. If that is the case, *our* fight in the Saar is simply *lost* and the *pastors* have *won* the game."[119] Indeed, local industrialists and their allies in the National Liberal Party leadership defined their party meetings during the spring of 1896 as "public" demonstrations designed for the purpose of responding to the alleged falsehoods of the reformist opposition, which had unsettled "public opinion" and threatened to turn it against the well-intentioned policies of the industrialist faction of the party.[120]

This conflict over social reform, the competition over the definitions and the sympathies of the local citizenry or public, and the struggle over the boundaries of permissible speech threatened to weaken the National Liberal Party during the Reichstag elections of 1898. The reform faction, now coalesced into the "liberal citizens' party," held a meeting at the end of May 1897 in order to criticize the current National Liberal leadership and its support for the Anti-Revolution Bill and for a bill proposed before the Prussian Landtag and calling for restrictions on the law of association.[121] These "threat[s] to public peace" and expressions of "unfounded mistrust of the maturity and independence of our people" were denounced, and the opposition put forward its own candidate of liberal reform, the jurist Wilhelm Kulemann of Braunschweig, whose platform committed him to a broad program of workers' social reform, "universal, equal, and secret suffrage," "equal rights," and state support for small urban and rural producers.[122] Kulemann soon withdrew himself from consideration, after many reform liberals expressed concerns that he was unknown to most voters in the region. After approaching the sociologist Max Weber, who declined the offer to run for the seat, reform liberals turned to Dr. Neff, the mayor of St. Johann, who was promptly elected as the new reform candidate in late January 1898. Two weeks later, when the paternalist wing chose Heinrich Boltz to run, the National Liberal Party had two candidates.[123] Throughout April, intense negotiations took place between party leaders around Boltz, Rhine Province party chairman Dr. Johannes, and Neff. In conciliatory tones, Neff stressed the need for the party to take a "liberal position," the acid test of which was the secret and direct franchise: "A liberal man of the present," he argued, "must also be socially minded, and the same rights must be good for all, including the right to vote for all workers."[124] But, fearful of a debilitating split within the party and the growing strength of the Center, Neff was convinced to withdraw in favor of Boltz in late April.[125]

Though this solution temporarily resolved the crisis, it also prompted the reformers and their left-liberal allies in St. Johann and Saarbrücken to seek a new candidate and, at least publicly, forced minor concessions from the industrialists. The reform opposition, now led by Dr. Meyer of the Artisans' Association, approached former commerce minister Hans von Berlepsch, who declined the offer to stand as the liberal candidate but encouraged opposition leaders to put forward their own candidate nonetheless. They chose Meyer, who announced a reform-liberal political program, which included the elements of Neff's earlier platform and stressed the associational rights of workers.[126] Meanwhile, as part of Neff's withdrawal, Boltz offered public guarantees of the "liberal" commitments of the National Liberal Party. Under the coordinating influence of Dr. Johannes, Neff and Boltz attended a 6 June campaign meeting, during which Boltz emphasized the similarities between his positions and Neff's program and reaffirmed his commitment to patriotism, the franchise, the right of association, and "liberal" extension of labor and social welfare legislation "under consideration of the practical interests of economic life and the maintenance of the ability to compete economically with other countries"—that is, without impinging on the interests of local heavy industry.[127]

The election in June brought unity between the reform and paternalist factions but also demonstrated the growing power of the Center Party in the central electoral district. Boltz easily outpaced Meyer in the first round, with 11,094 to 2,198 votes; Meyer then urged his supporters to vote for Boltz in the second round.[128] The most effective source of party unity, however, was the dramatic increase in the number of Center votes. The more concentrated efforts to mobilize workers by means of the Catholic miners' and factory workers' associations and the local Volksverein during the electoral campaign itself allowed Center candidate Euler to emerge with 10,806 votes in the first round of voting twice the number of votes the party received in the previous Reichstag election. This success was the first demonstration of the party's ability to actually capture the seat. The Center campaign focused explicitly on social reform and the right of workers to organize, while also stressing the importance of defending the artisan or small producer. The campaign was led by the priest Becker, Dr. Strauss, and Matthias Oesterling, leading figures in the Volksverein and the Saar League of Catholic Miners' and Factory Workers' Associations. The small group of local Social Democratic Party supporters, who ran Leopold Emmel in the first round of the election, urged support for the Center candidate in the second round.[129]

Nevertheless, the victory of the Boltz and Saar industrialists in both the internal National Liberal Party struggle and the elections of 1898 could not obscure the ways in which a regionwide conflict over the shape of labor relations in Saar heavy industry brought new challenges to the "Stumm system." These challenges began with the state *Sozialpolitik* of the New Course and the self-mobilization of Saar miners and other workers from 1889 to 1893, which dramatically called into question the structuring parameters of local *Öffentlichkeit.* After the miners' defeat, the Catholic and Protestant-liberal reform movements continued to contest paternalist control over the main discourses of local *Öffentlichkeit*—a challenge that produced a crisis within the ruling paternalist bloc from 1895 to 1898, which divided nearly all of the institutions of the local Protestant *Bürgertum:* the Saarbrücker Casino, the local gymnasium and *Oberrealschule,* the district synod, the leading newspaper (*Saarbrücker Zeitung*), and the National Liberal Party organization. The movements also challenged the hegemonic ideological discourse of employer paternalism itself.

CHAPTER 4

Workers' Associations, the Struggle over *Öffentlichkeit,* and the Crisis of Paternalism

The "War between the Patriots" exposed and deepened the fissures within the ruling social coalition and the dominant party-political cartel, but the dramatic expansion of working-class associations and labor organizations after the turn of the century brought the first systematic challenges to paternalist authority in the Saar. During the decade before 1914, Saar industrial workers began to form their own leisure clubs, civic associations, and occupational organizations in exponentially increasing numbers, in ways that brought working-class men and women as independent actors directly into the routines and domains of the local public sphere. These increases took place in the sociologically more heterogeneous cities of Saarbrücken and St. Johann, but they were much more dramatic in the industrial towns of Malstatt-Burbach, Neunkirchen, Sulzbach, and Dudweiler and in smaller mining communities like Illingen, where the number of associations rose from three to over fifty from 1890 to 1914.[1] This rapid proliferation of independent workers' organizations energized local social life with new interests and constituencies, and working-class civic associations even disrupted the comfortable certainties of notable-driven municipal politics throughout the region—developments that alarmed local officials like Saarbrücken *Landrat* Bötticher, who noted the difficulties of keeping track of working-class and especially socialist meetings and organizations owing to sheer numbers. The "insufficiency of police personnel," he pointed out in 1904, made "surveillance of the almost daily meetings and investigation into the actual kind" of associations holding such meetings nearly impossible.[2]

The transformation of the public sphere after 1900 was linked to the emergence of labor movements in the Saar, which Karl Rohe has described as a "delayed region," where the preeminence of Christian unionism over socialism and the relative "lateness" of independent labor mobilization were unique in relation to developments in other industrial districts and regions, including Berlin, Hamburg, Saxony, and even the "late"-developing Ruhr.[3] From 1899 to 1903, Social Democrats began to build a fledgling trade union movement and party political organization among industrial workers in the region. This activity and concern about the impending formation of interconfessional, antisocialist trade unions prompted Catholic clergymen and laymen to create a network of antisocialist and conservative Catholic workers' associations, with their own trade sections (*Fachabteilungen*) for specific categories of workers, beginning in 1902. But the latter were not able to halt the advance of the biconfessional (though mostly Catholic) Christian trade unions, which were established in 1904 and became the leading representatives of Saar industrial workers, especially miners. The relative success of the Christian trade unions over the Social Democratic unions, according to Klaus-Michael Mallmann, was a result of the distinctive confessional geography of the Saar, where the confessionalization of labor conflict, pitting Catholic proletarians against Protestant employers and managers, increased the appeal of a religiously inspired Christian unionism among Catholic migrants.[4]

This salutary perspective acknowledges the importance of religion in the formation of workers' political subjectivities, but it tends to obscure the varying adherence to Christian unionism among Saar wage earners and the structuring parameters and complex workings of the public sphere, as well as the specific political modalities of its transformation after 1900.[5] To understand the relative success of the Christian trade unions and the dynamics and consequences of labor mobilization in the prewar Saar, it is important to look beyond milieu-based considerations of confessional geography and toward the specific constellation of discursive relations and institutions, including the state, which determined access to the public sphere. The public sphere in the Saar was not a uniform, even, or plural social-discursive space but, rather, an ideological field structured in relations of social inequality and discursive hierarchy.[6] It allowed certain subordinate or oppositional discourses, such as social Catholicism and Christian unionism, and disallowed others (namely, social democracy). It contributed to the distinctive urban-industrial and confessional geography of the regional labor movements—that is, the (albeit limited) success of Social Democratic organizations among artisans and skilled workers in medium-sized or small businesses located in the larger, predominantly Protestant cities and towns versus the success of Catholic trade unions in smaller mining communities and villages. Nevertheless, within these confines, the Saar labor movements began to challenge long-standing paternalist practices and institutions in the region. This chapter explores the ways in which workers, Catholic labor leaders, Christian unionists, and Social Democrats struggled over the very definitions and conditions of *Öffentlichkeit* and gradually undermined the structures of paternalist public authority in the Saar after 1900.

Workers, Associational Life, and Municipal Politics

The expansion of working-class associational life challenged the established routines of paternalist *Öffentlichkeit* in what was once a safe bastion of employer influence: municipal government. Similar to developments elsewhere in Germany, the scope of local administration and public services throughout the Saar region expanded exponentially during the years after 1890. The construction of a whole host of new utility projects (especially gas, electricity, and water services) and transportation facilities (including streets and rail lines and connections) and the expansion of existing facilities dramatically transformed the physical spaces and infrastructure of Saar municipalities and communes—a transformation most visible with the unification of Saarbrücken, St. Johann, and Malstatt-Burbach into the *Grossstadt* of Saarbrücken in 1909. These changes resulted in larger municipal councils or at least discussions about the need to increase the number of councillors; the rapid proliferation of new municipal committees and agencies; and, above all, the dramatic increase in the numbers of civil servants and municipal bureaucrats, who were necessary for the administration of city services. They were financed in large part by increasing tax revenues levied not just on the incomes of local industrialists and notables but also on the rising incomes of small business owners, artisans, and industrial workers, especially miners in the local mining communities. In this context, third-class and even some second-class voters began to form new civic associations (*Bürgervereine*) that challenged the monopoly of industrialist notables over the town councils after 1900.

Modeled on existing bourgeois organizations in the sociologically more heterogeneous cities of Saarbrücken and St. Johann from the 1860s to the 1880s, these organizations were run by varying coalitions of shop owners, artisans, and industrial workers but reported largely working-class memberships. Adopting such names as Local Interest Association, Association for the Discussion of Municipal Matters, Association for the Protection of Local Interests, or simply Citizens' Association, they proliferated rapidly after 1905 throughout towns and communities as diverse as Malstatt-Burbach, Dillingen, Altenwald-Hühnerfeld, Sulzbach, Illingen-Gennweiler, Merschweiler, Altenkessel-Neudorf, Herrensohr, Fischbach, Hüttigweiler, Uchtelfangen, Wemmetsweiler, Wiebelskirchen, and Russhütte. In the mining communities, miners dominated or even led the new associations, which claimed much larger memberships and evinced even higher levels of political mobilization than the associations in the medium-sized industrial towns. The effort to form the *Bürgerverein* of Illingen, for example, was chaired by the carter Peter Klein and the activist miner Johann Schäfer.[7] The most impressive organizational success of this kind was achieved by the Association for the Protection of Local Interests in Altenwald-Hühnerfeld, predominantly a mining area that was also the site of the Röchling coking plant. By the fall of 1908, its chairman, house-painter Heinrich Michael Losch, presided over an organization of 668 members, of whom 425 were active miners, 55 were pensioned miners, and the remaining 188 were nearly all wage earners, white-collar workers, and artisans.[8]

Together, the new *Bürgervereine* constituted a novel and broadly based social movement throughout the region—an emergent proletarian public, with its own organizations and means of public communication, which demanded more "openness" (*Öffentlichkeit*) and accountability in municipal affairs. The numerous associations were supported by activists like Johann Schäfer, who encouraged the formation of *Bürgervereine* and delivered the "most bombastic speeches" at different meetings throughout the district of Ottweiler; indeed, local officials believed his aim was to "unify all *Bürgervereine* in one larger movement."[9] They were also brought into a shared field of discourse by the efforts of Hermann Josef Meurer, who moved to the Saar from Essen-Altendorf and published a regional newspaper called *Citizen's Friend* (*Bürgerfreund*), which became a central medium of this emergent public.[10] Local activists self-consciously conceived of the civic associations as public organizations in which Saar "citizens" could register complaints and file petitions in matters related to the expropriation of property for public building projects, the financial costs of street and building construction, canalization, the leasing and unequal use of public facilities (e.g., gas and waterworks, electricity, sanitation, and the rail system), the enforcement of housing codes, the size and purpose of municipal bureaucracies, and taxes.[11] But their principal goal was to mobilize local residents in the cause of municipal self-governance, especially during election time. According to the editors of *Citizen's Friend,* "The heretofore peaceful, much too peaceful citizen [*Bürger*] would like more energetically to defend himself, to engage in his own behalf. The municipal council elections provide the opportunity to do this. . . . These election days can and should be accountability days [*Zahltage*], during which deceitful municipal councillors who are unfaithful and lose the trust of their voters are kicked out."[12] In the process, they challenged the definitions of *Öffentlichkeit* as a criterion of openness and its realization in the work of Saar municipal councils.[13]

Already in 1901, the municipal council in Neunkirchen entertained its first motion regarding "the principle of *Öffentlichkeit,*" in response to petitions from local citizens and the "active desire of a large portion of the local residents for openness [*Öffentlichkeit*]"; as a consequence, it gave three different newspapers access to its meetings.[14] In October 1909, the failure of the local council to publicize its agenda moved editors of the *Völklinger Zeitung* to insist on the broader relevance of council affairs and to question the failure of the councillors to publicize their activities in the steel town of Völklingen.

> One asks involuntarily: Why? Why this petty-minded secrecy? After all, this exclusion of the public [*Öffentlichkeit*] may only take place for the *most serious of reasons.* We believe that the still legally valid though obsolete Rhine Municipal Ordinance, which is completely unsuited to our industrial city, should not serve as a convenient means by which to keep the citizens in ignorance and doubt about their most important public issues.[15]

In 1911, Chairman Bößmann of the Dudweiler association called for a municipal ordinance that was "appropriate to the times" in a way that "set in law" the "duties of provincial officials" and the "openness" (*Öffentlichkeit*) of the municipal council meetings.[16] In these ways, the new civic associations, assisted by some local newspapers and even some sympathetic councillors, attempted to expand prevailing definitions of *Öffentlichkeit* by stressing the importance of open discussion of town affairs, regular press reporting of the minutes of council meetings, "public control over . . . council meetings," or the role of civic associations as a "connective channel" between the council and the wider citizenry.[17]

Members of the new *Bürgervereine* articulated their formal aims in general appeals to citizenship and public accountability, but they also drew on a latent class-oriented perspective that stressed the unequal economic burdens imposed by municipal growth and railed against the overriding influence of heavy industry, in ways that challenged the paternalist structures of local *Öffentlichkeit* as a collective subject. They routinely invoked the "public," the "public interest," the "public well-being," and the wider "citizenry" (*Bürgerschaft*) in order to stake their own claims and legitimize their concerns, and they claimed to speak on behalf of "all occupations and classes of citizens," especially the "tenacious farmer," the "industrious artisan," the "active small businessman," and the "struggling industrial worker."[18] But civic activists like Meurer, who were often connected to the Christian and Social Democratic unions, also emphasized the importance of the labor movements for galvanizing the sentiments of workers and "other occupational groups" in favor of "political freedom." They defined the principal lines of municipal conflict in class terms by claiming to represent the taxpaying citizenry or "citizen's party" (*Bürgerpartei*) in opposition to the local ruling coalitions of mining officials, factory owners, and managers—the *Hüttenpartei* or *Grubenpartei*—which dominated local councils. In Illingen, association members regularly complained of the perceived inequalities in water usage: local heavy industry, *Bürgerverein* chairman Peter Klein argued, received more than their share and a better quality of water than local residents.[19] Chairman Peter Beckinger of the Dillingen *Bürgerverein* criticized plans of the Dillingen steelworks to obliterate a much-used thoroughfare as part of its plans to expand its factory grounds, and he rejected the notion that local citizens were "entirely dependent on the factory."[20] In Dudweiler, the miner Peter Henrikus, a member of the Local Interests Association, argued forcefully that Dudweiler councillors should begin to reflect the sociology of the population at large; since fully three-quarters of Dudweiler's residents were workers, he claimed in 1911, then fifteen of the council seats should be occupied by workers and the remaining nine by officials and businessmen (*Beamte* and *Geschäftsleute*).[21] Henrikus, like many other civic association activists, thereby directly challenged the long-standing paternalist calculus governing the distribution of municipal council seats throughout the region.

Nevertheless, the desire to avoid overtly religious and political themes in association statutes and during meetings and the tendency to embrace the central discursive strategies of bourgeois *Öffentlichkeit* muted the activism of the *Bürgervereine* and curbed their potential militancy. As they mobilized citizens (*Bürger*) against officials (*Beamten*) of local mining inspections or factories in an effort to claim the universal interest of the public against the particular interest of capital, the associations' claims to universality could come undone in the citizen-worker distinction. This happened in the Dudweiler *Ortsinteressenverein* [Local Interests Association] in 1911, when association chairman Bößmann, the director of the local mining dormitory, argued in favor of a more equitable distribution of council seats (eight for mining officials, seven for local small businessmen, and nine for workers) but also warned workers "not to allow themselves to be incited" and "to vote for people who will represent the interests of the citizens." The existing forms of representation should be maintained, Bößmann argued: "The worker is not in a position to represent the interests of the citizens—labor makes him tired and weary; they [*sic*] should elect people around them (even mining officials) who have the interests of the workers in view and in whom they can place their trust."[22] From Bößmann's perspective, industrial workers did not qualify as universal citizens capable of representing the interests of the "public."

More serious was the threat from employers and local state officials, who grew concerned about the political implications of this bourgeoning civic activism, the growing "politicization" of municipal council elections, and the fracturing of the putatively unitary "public." Already in 1903, mining officials were sacking workers who dared to stand for elections to local councils. In Püttlingen, for example, the miner Edlinger was dismissed for his involvement with an electoral committee composed of miners, who planned to support Edlinger as a candidate and thereby challenge the numerical preponderance of mining officials in the local *Gemeinderat* [town council].[23] Employers in private heavy industry were even more thorough. Officials of the Dillingen steelworks, who forced the resignation of a popular mayor of Dillingen in the fall of 1903, sent company employees to the meetings of the newly formed *Ortsinteressen-Verein* in January 1908 in order to compile lists of any workers in their employ who might be attending; made sure that copies of all association petitions with member signatures were distributed to all of the councillors on the Dillingen council, who of course included officials of the steelworks; and pressured their employees to stay away from the organization. By February 1908, the sudden exodus of most of its members had practically destroyed the civic association in Dillingen.[24] At the same time, local state officials proceeded in more legalistic ways. In January 1907, *Landrat* Laur of Ottweiler notified all mayors in his district of the "unwanted" involvement of citizens' associations in local elections and public assemblies. Insisting they provided a forum only for the "machinations" of self-interested individuals and therefore not the general concerns of the "public," he urged the mayors to maintain close surveillance over civic association meetings; to proceed "severely" (*unnachsichtlich*) with criminal charges against any person who slanders a public official; and to resist conferring corporation status on any association, in order to head off attempts to gain the legal right of petition.[25] These activities were made easier by a January 1908 decision of the Saarbrücken Criminal Court, in a case involving efforts to suppress the activities of women in the Dillingen citizens' association, which determined that the civic associations were to be treated as political organizations and subjected to regulations regarding the official registration of meetings and police surveillance.[26] Finally, in the rare instances when association members spoke in ways that violated the norms of polite speech, as Johann Schäfer did in Illingen in March 1907, officials promptly filed criminal charges against them.[27]

These actions of industrialists and state officials and the self-limitations of some civic organizers put the brakes on the broader movement of civic associations by 1909–10. The last issues of *Citizen's Friend* were apparently published in the fall of 1909 in Saarbrücken and in April 1910 in Neunkirchen; and by January 1910, the police in Saarbrücken were reporting that the paper could no longer attract subscribers, though the civic associations themselves were not entirely dissolved.[28] Nevertheless, the *Bürgervereine* did succeed in bringing new working-class constituencies into the arena of municipal politics in ways that challenged the paternalist hegemony of industrialist notables in the local public sphere. Their protests threatened to open up municipal council meetings and paternalist priorities to broader scrutiny and raised issues of municipal accountability that could no longer be quietly ignored or easily silenced, especially as their activism merged with the Saar labor movements after 1906.

Social Democrats, the State, and the Public Sphere

The general proliferation of working-class associations was both the precondition and the consequence of trade union mobilization, including the efforts of Social Democrats, in the Saar. From their bases in Saarbrücken and St. Johann, where they had established an electoral association in 1898 and a trade union cartel in 1899, Social Democrats built the embryonic institutions of a socialist public in the region after 1900. This activity began with discussions about organizing Saar miners (which took place as early as 1899 within the national leadership of the Free Trade Unions) and with increasing numbers of meetings held by local socialists in 1901 and 1902. But concerted efforts to break into the "Kingdom of Stumm," led by Nikolaus Osterroth, a clay miner from the Palatinate, were launched in 1903, and the main recruitment drive began on 1 May 1904.[29] Local union organizers also set up a workers' secretariat for dispensing legal aid to workers in St. Johann; and in December 1904 and January 1905, they began to publish the first issues of the Social Democratic newspaper *Saarwacht*. By these means, Saar Social Democrats generated their first effective public presence throughout the region: though their largest political assemblies still took place in Saarbrücken and "red St. Johann," they were also able to hold their first public meetings in the industrial towns of Völklingen, Malstatt-Burbach, and Neunkirchen after 1905. In February 1907, they even established an association for the district of Ottweiler in Neunkirchen, a location the bricklayer and civic association trustee Johann Lohr called the "blackest corner of Saarabia."[30] By the summer of 1907, Otto Hué, national chairman of the socialist miners' union, the Alter Verband, was speaking to audiences of five hundred people in Neunkirchen; and by 1912, the Alter Verband was holding mass meetings in the small industrial village of Wiebelskirchen.[31]

Nevertheless, the Social Democratic movement in the prewar Saar never quite broke out of the urban-industrial geography of paternalism and the constraints of the local public sphere. At the end of the "Stumm era," Social Democratic supporters were employed almost exclusively outside the reach of heavy industry—mostly in the small workshops, construction sites, and state-owned rail yards in Saarbrücken and St. Johann—and in areas of Protestant or confessionally mixed settlement. After 1900, moreover, the SPD's electoral association for the district of Saarbrücken never moved much beyond its sociology and location of 1898, when it comprised mostly joiners and masons from Saarbrücken and St. Johann, even though Social Democrats did manage to attract increasing numbers of supporters in Neunkirchen and in heavily Protestant Wiebelskirchen, a community of miners after 1907.[32] In addition, the Free Trade Unions, which were only able to establish local branches among tailors, joiners and shoemakers, construction workers, painters, printers, and bakers during the 1890s in Saarbrücken and St. Johann, failed to move into most of the industrial towns and mining villages of the region after 1900.[33] In 1902, they reported only ten local branches with 513 members, none of whom were miners or steelworkers; and by 1908, after five more years of organizing, they could report only twenty-eight local branches, still mostly based in the construction, artisanal, and metalworking trades, with 3,835 members. By the end of 1910, the Alter Verband claimed only 862 members from among the 50,802 miners employed in the state-run Saar mines.[34] Not surprisingly, Social Democratic Party organizations, which managed to attract only 777 members by 1913, faired poorly in regional Reichstag elections. When it won a third of the popular vote nationally and sent more delegates to the Reichstag than any other party in 1912, the SPD was able to win only 7.8 percent of the vote in the electoral district of Saarbrücken, 4 percent in Ottweiler, and 3.7 percent in Saarlouis.[35]

These patterns of Social Democratic mobilization and the limited appeal of socialism among workers in the prewar Saar were shaped by the ideological forces and institutional barriers that determined access to the local public sphere. The latter included an already existing and growing Catholic public actively constructed in opposition to the threat of socialism, with a long-standing network of religious leadership and associations, new occupational associations (especially the Catholic workers' associations and Christian trade unions discussed later in this chapter), and Center Party organizations.[36] The institutional authority of local priests and teachers, from the pulpit and in the schools, reinforced this opposition. Catholic priests, like many Protestant pastors in the Saar, made no secret of the fact that they routinely took an opportunity during masses to warn their parishioners away from social democracy—a practice Saar miners revealed in a survey (which kept names confidential) conducted by the social commentator Adolf Levenstein in 1912.[37]

More serious was the fierce resistance of the leading Saar industrialists, who closely monitored socialist activity throughout the region and coordinated a series of antisocialist measures with the assistance of state officials. In 1903, employers and managers from all of the leading local firms posted notices on their factory gates in response to the Social Democratic recruitment drive and threatened immediate dismissal of workers who participated in socialist activities.[38] In January 1905, officials of the Neunkirchen, Röchling, Burbach, Halberg, and Dillingen steelworks issued similar warnings with the appearance of *Saarwacht*.[39] Moreover, in

December 1903, in order to create a comprehensive system of surveillance and blacklisting, Chairman Ewald Hilger of the Mining Office met with the directors of nearly all of the principal mining and steel concerns located in Völklingen and Hostenbach, Malstatt-Burbach, Brebach, Neunkirchen, St. Ingbert, and neighboring Lorraine and the Palatinate. Run by the Mining Office in Saarbrücken, the system entailed the organized exchange of information about trade union meetings and activities, with the circulation of reports to all participating employers as well as to the *Landräte* in Saarbrücken and Ottweiler, the *Kreisdirektor* [county director] in Forbach (Lorraine), and the director of the railroad administration in St. Johann. In accordance with this arrangement, company officials and informants attended local union and socialist meetings, reported on the content of each discussion, and identified the workers present so that their names could be sent to their employers.[40] By the spring of 1904, Saar industrialists had established their own newsletter, the *Gewerkschaftliche Nachrichten* (Trade Union News), which reported regularly on all socialist and trade union activities and, by the fall of 1904, circulated blacklists compiled by local firms under a common policy of sacking all workers who attended SPD or trade union meetings, read socialist literature, or in any way expressed socialist sympathies.[41] This kind of surveillance and intimidation and their consequences for workers—which ranged from the sack or transfer to less remunerative work to physical attacks and, in one case, "hunt[ing]" the Social Democrat Osterroth out of town with a fire hose—amounted to a wide-ranging effort to police the expanding spaces of the local public sphere, from the factory workplaces and mines to the local taverns and streets, and directly involved state officials in the long-standing tradition of local collusion.[42]

Nevertheless, Saar Social Democrats introduced a new style of political mobilization, which directly and indirectly—insofar as it forced others to contend with Social Democratic methods—expanded the sites and the constituencies of *Öffentlichkeit* in the region. Local organizers launched a grassroots effort to distribute Social Democratic leaflets, brochures, and newspapers outside the factories and mines, in the workers' dormitories, on trains, and from house to house—often delivering them at night to avoid the authorities.[43] These efforts initially began to bear fruit: Social Democrats began to attract supporters in the industrial districts and even in Catholic communities long closed to their organizations. Despite the difficulties of identifying socialist sympathies in a highly repressive environment like the Saar, it is clear that increasing numbers of Saar steelworkers, miners, and glassworkers, a number of whom were Catholic, began to read socialist literature and attend meetings of the Free Trade Unions after 1903. These included employees of the Röchling steelworks in Völklingen, the Burbach steelworks and the Dingler & Karcher machine factory in Malstatt-Burbach, the Dillingen steelworks, and the Reppert glassworks in Friedrichsthal.[44] Saar miners demonstrated even greater interest. The Social Democrat Karl Krämer found a growing number of receptive miners in 1903–4 and was preparing to set up a branch of the Alter Verband in Sulzbach; when local officials searched his residence in the spring of 1904, they found a list of eighty Saar miners who were receiving the Alter Verband's newspaper. By 1905, some 280 miners had subscribed to the paper, and by September 1912, according to the *Landrat* of Ottweiler, the number of Saar miners who belonged to the socialist Alter Verband exceeded one thousand.[45] Through efforts at grassroots mobilization beyond the meeting hall, therefore, workers in smaller communities, such as Püttlingen and Wiebelskirchen, and in the larger towns of Sulzbach and Neunkirchen (once the central redoubt of Stumm's paternalism) had opportunities to become involved in union and political matters.

With similar consequences, Social Democratic mobilization addressed women and entire families in direct ways. Despite their privileging of class over gender and periodic references to the male breadwinner, Social Democrats formally embraced sexual equality in their party program and solicited women's active participation in trade union and political activities more than all other organized labor and political movements in imperial Germany. In the Saar, women (and children) participated in leafleting activities in the house-to-house campaigns, but female party members were also involved in these efforts in the areas around the main factories. In 1908, for example, local party chairman Emil Leimpeters was distributing handbills in Burbach with two female party "comrades" (*Genossinnen*), Frau Ruff and Frau Wolz, both of whom were physically threatened by officials of the Burbach steelworks.[46] Moreover, unlike leaders of other political parties and labor organizations, Social Democratic trade union and party activists encouraged women to attend their meetings, both as wives of workers and as workers themselves. As early as March 1903, twenty-one out of forty-nine participants in a metalworkers' meeting in St. Johann were women; in April 1903, women were attending meetings of glassworkers in St. Johann; and in August 1903, women were invited to attend the first major miners' meeting held by the SPD in St. Ingbert. This practice continued after 1908, when the Prussian law of association was changed to allow for the participation of women in political meetings and organizations. By 1910, at least one local woman, Frau Pflüger, was a leader of the party organization in Neunkirchen, and women were routinely attending meetings in Neunkirchen and Wiebelskirchen.[47]

These forms of mobilization were part of a more militant style of activism, often linked to actions on the shop floor, during the period from 1903 to 1906. The first workers to organize labor actions and conduct strikes were the Social Democratic construction workers, brewery workers, and printers in Saarbrücken and St. Johann. By the spring of 1904, Social Democratic printers had successfully organized workers employed by all of the major publishers in Saarbrücken and St. Johann, had secured collective wage agreements, and were attempting to win recruits from among the employees of August Spiess Press, an industry-friendly publisher in Malstatt-Burbach. In early 1904, the local trade union cartel, led by the typesetter Christmann, carried out a successful boycott of the Mügel Brewery, which secured higher wages and the rehiring of a dismissed activist. It shortly thereafter launched a similarly successful action against the Aktien Brewery of St. Johann-Saarbrücken. At the same time, 354 bricklayers launched a three-week strike against local construction companies in May 1904.[48] These actions were followed by socialist attempts to move closer to the centers of Saar paternalism, when employees of the Reppert glassworks in Friedrichsthal established a small local branch of the Social Democratic Union of Glassworkers and Related Trades (Verband der Glasarbeiter und Arbeiterinnen und verwandter Gewerke Deutschlands) in 1901. In March 1906, they launched a successful strike against the company in response to declining wages and inadequate raw materials.[49]

Another Social Democratic strategic innovation lay in taking advantage of the state-mandated industrial courts and sickness insurance funds, which included both workers and employers on their governing boards, in order to expand the boundaries of the local public sphere.[50] Like the pattern of socialist mobilization more generally, political contestation over the composition of the industrial courts and the boards of the sickness funds followed distinctive urban-industrial patterns and benefited mostly workers from the artisanal trades in the larger towns, small businesses, and the smaller machine and metalworking factories in the Saar. As early as 1902, Social Democrats managed to elect a majority of worker representatives to the industrial court in Saarbrücken.[51] By 1910, they had secured eight out of twelve mandates in the second chamber, responsible for disputes within small industry; but in the first chamber, with jurisdiction over workers in heavy industry, they did not even put up a list.[52] In the elections to the boards of the sickness funds, in which Social Democrats were able to establish a foothold, a similar pattern is evident. They gained control over the governing boards in Saarbrücken and St. Johann by 1903; indeed, most of the employer representatives in these institutions were socialists (three or four of six in Saarbrücken and four of six in St. Johann), and in St. Johann, most of the delegates to the general assembly were socialists (104 out of 116). By contrast, in Malstatt-Burbach in 1903, there were no socialist members on the board of the *Ortskrankenkasse* [local sickness insurance fund], and there was no organized trade union to even put up a list.[53] When the socialist unions finally did manage to contest the elections in Malstatt-Burbach in December 1906, they ran up against a unified employer opposition, were expelled by the chairman after their election to the board, and then were defeated by a coalition of employers, confessional workers' associations, and Christian trade unions in a subsequent election.[54] Nevertheless, by 1914, the socialist unions had captured a majority of the workers' seats on the governing boards of the sickness funds in Saarbrücken and St. Johann and the umbrella sickness fund created by the unification of Saarbrücken, St. Johann, and Malstatt-Burbach into one *Grossstadt* in 1909; they had also won a majority of workers' seats in Neunkirchen; and they were contesting elections in Völklingen.[55]

This "legalistic" strategy of turning state institutions to their own ends was organized through a discourse of class interest and citizenship rights, which avoided explicit invocations of *Öffentlichkeit* as the basis on which to make political claims. Saar Social Democrats generally spoke in the name of workers and their class interests, rather than overtly in the name of the "public" interest, in a language of class emancipation that emphasized connections between economic privation and exploitation and the infantilizing effects of company paternalism.[56] In response, however, to formidable efforts to suppress their movement, Saar Social Democrats adopted a political strategy focused on the struggle to defend the worker's legally guaranteed right to organize—that is, a strategy of gradualist proceduralism and constitutional legality. In their appeals to workers, they equated work relations and the paternalist "system of cultivation" with "slavery" and the "deprivation of rights." The local Social Democrat Karl Krämer, in an orientalist formulation common to radical and liberal critiques of the day, likened the "systematic repression of workers' freedoms" in the Saar to conditions obtaining in an "Asiatic" state. In "Saarabia," he maintained, workers were treated as "underage children" or "subjects" and denied their legal right to association and equality before the law.

> "All citizens are equal before the law" is the meaning of the fundamental principle of every constitutional state. Through the Imperial Industrial Code (section 152), we miners are also granted the right of free assembly, which allows us to meet in private organizations to discuss grievances from work and to obtain better wages, etc. No one has the right to deny us these organizations of the basis of the industrial code. Nevertheless, there exists in Saar mining no trade union organization, because each comrade knows or assumes, on the basis of experience, that he will suffer "economic disadvantages" at work if he makes use of his legal right to association in a way deemed unacceptable by "those above."

In this way, local Social Democrats linked economic deprivation and the lack of a "humane existence" to the systematic subversion of workers' "constitutionally guaranteed citizenship rights" by employers and called on industrial workers to "wake up," to "break the chains of slavery," and to make use of their right to organize.[57]

This strategy had important consequences for *Öffentlichkeit* in the region, because local state officials, as "neutral" civil servants, were vulnerable to charges of violating workers' constitutional rights—a vulnerability dramatically revealed in a very public libel trial involving the prosecution of Saar miner and Social Democrat Karl Krämer and initiated by Chairman Ewald Hilger of the Mining Office in 1904.[58] Sacked for his attendance at a socialist meeting in St. Ingbert in August 1903, Krämer was prosecuted for distributing handbills (cited in the preceding paragraph) that criticized the "oppressive" paternalist labor regime in Saar mining, for its low wages and its restrictions on the worker's "legal right to association."[59] The eight-day trial in Saarbrücken ended in Krämer's conviction (later dismissed by an appeals court), but not before his lawyer, Social Democrat Wolfgang Heine, was able to bring sixty-nine witnesses to the stand and thereby generate a wider public discussion of labor relations in Saar mining. More than any other local event, the trial brought attention to the low wages and poor living conditions of Saar miners, the frequent verbal harassment and corporal punishment meted out by *Steiger,* abuses committed by *Knappschaft* doctors, and the extensive electoral influence brought to bear on Catholic and Social Democratic voters by the Mining Office. In the end, it exposed the antisocialist and anti-Center practices of the Mining Office to both regional and national publics. Both Hilger's unabashed defense of the policy of sacking all workers linked in any way to socialist activities and the widespread evidence of discrimination and intimidation of Catholic miners during elections angered not only Social Democratic, liberal, and Center politicians and supporters but state ministers and officials as well. Concern over negative publicity prompted Hilger's resignation as chairman of the Mining Office in February 1905.

In response to criticism and out of fear that deploying long-standing paternalist measures of coercion could ultimately backfire and generate support for the Alter Verband, Hilger's replacement, Wilhelm Cleff, implemented a new antisocialist strategy that redefined the boundary between private and public in relation to the political activities of Saar miners. The Mining Office would no longer dismiss miners for their "passive" involvement in socialist activities—that is, for attending meetings, reading socialist newspapers, or voting for the SPD. Instead, it would sack only the "leading personalities" of the movement and miners who engaged in "public agitation for social democracy." This meant that Social Democratic miners were sacked for chairing or renting halls for Alter Verband or SPD meetings, speaking during the latter, or distributing socialist leaflets. Moreover, in a policy change of enormous consequences, officials of the Mining Office decided they would no longer sack or punish miners for any kind of involvement in the Christian Miners' Union of Germany (Gewerkverein christlicher Bergarbeiter Deutschlands).[60] The aim was to contain the Social Democratic movement by preventing any employed miners from becoming leaders of the trade union or party organizations. This new policy, mining officials believed, would achieve its aim in several ways: it would reduce the number of local functionaries and burden party finances by forcing its members to support all of their leaders financially; remove Social Democratic leaders from the pits and prevent their recruitment efforts at the workplace and their involvement in works council and safety steward elections; and limit the appeal of local socialist organizers to those still employed. Above all, according to Cleff, the new policy made it possible to keep a better eye on the Social Democratic movement, since the practice of dismissing workers for any and all SPD affiliations would force socialist miners underground—in which case, Cleff argued, "all too many embers would gather beneath the ashes, and we could be surprised by a sudden explosion, the scale of which we would be unable to measure in advance."[61] In July 1910, when mining management discussed the possibility of renewing contacts with local industrialists for the purpose of combating social democracy with all available means, Cleff explicitly acknowledged the differences between local private industry and the state-run mines that motivated this policy: "If the private employer is allowed to use any means which the law and his own interest permits in this struggle," Cleff argued, "so the state as employer is enjoined from adopting measures that contradict the demands that public opinion makes on the state administration in regard to its labor policy."[62] Against the backdrop of a series of state policies and constitutional guarantees, including the passage of a new law maintaining the secrecy of the Reichstag ballot in 1903 and the imperial government's decision not to seek coercive legislation to curb the growth of social democracy after the failure of the Penitentiary Bill of 1899, Cleff and the leadership of the Mining Office acknowledged the end of the "Stumm system" in Saar mining.

Catholic Labor, Christian Trade Unionism, and the Limitations of Bourgeois *Öffentlichkeit*

Saar Social Democrats realized few direct benefits from the actions of the Mining Office, but the confessional labor movements were directly privileged by the new policy. Christian labor in the Saar region was divided between the supporters of the Catholic workers' associations and the Christian trade unions. During the summer of 1902, the former set down roots in the Saar as members of the League of Catholic Workers' Associations, Headquarters Berlin (Verband der katholischen Arbeitervereine Deutschlands, Sitz Berlin), as part of the Catholic clergy's wider response to the rise of social democracy in the wake of *Rerum Novarum.* But the more immediate cause of concern was the growing consolidation of the Christian trade unions, which created an umbrella organization in the National League of Christian Trade Unions on 1 January 1901, and the appeal from the Catholic workers' associations in Cologne in 1899 encouraging Catholic workers to join the biconfessional unions. In response, in August 1900, Cardinal Georg Kopp of Breslau, with the support of Felix Korum, the bishop of Trier, issued the Fulda pastoral, which called on all priests and Catholic organizers to oppose the bi-confessional organizations and to build workers' associations, especially trade sections (*Fach-*

abteilungen) according to the branch of industry, under the auspices of the Catholic Church.[63] Concentrated in the dioceses of Breslau and Trier, which included the Saar region, the Catholic workers' associations of the "Berlin Direction" developed into an important antisocialist and conservative movement, which also opposed other Catholic organizations, including the League of Catholic Workers' Associations of Western Germany (Verband katholischer Arbeitervereine Westdeutschlands), the League of Catholic Workers' Associations of South Germany (Verband katholischer Arbeitervereine Süddeutschlands), and the Volksverein für das katholische Deutschland.

The formation of the regional organization for the Saar district came after a meeting of 122 priests on 26 June 1902 in Trier and began with the creation of a diocesan league the following November. Led by priests and executive committees staffed by local clergy and laypersons, the Catholic workers' associations and their trade sections offered a wide range of assistance to workers, including representation in matters related to work and wages; financial support during lockouts, strikes (launched by other workers), periods of unemployment, and cases involving disciplinary actions (*Massregelungen*) by employers; and legally nonbinding benefits, ranging from funeral and moving allowances to support in the case of sickness.[64] In addition, they sponsored a program of general sociability and cultural activities through their "entertainment evenings" and other social events, annual "social courses" for the purpose of "schooling" association leaders and "workers in contemporary social questions," and regional workers' secretariats that provided legal aid to Catholic workers in St. Johann and Neunkirchen.[65] On this basis, Saar priests and Catholic laypersons were able to build a sizable regional movement: after 1907, league numbers leveled off at eighty-six associations with eighty-two *Fachabteilungen* and 16,500 members, mostly miners.[66]

Catholic leaders defined the new workers' organizations as first and foremost antisocialist formations, but they also sought to counter the activities of the Christian trade unions—a position that dovetailed with employer paternalism. Whereas the Social Democrats were attacked for their materialist irreligion and their revolutionary challenges to the existing social and political order, the Christian unions were accused, in contradictory ways, of unnecessarily exposing Catholic workers to Protestantism within the biconfessional organizations and of encouraging secular views by their exclusion of religious considerations from the calculation of occupational interests.[67] This not only removed the authority of the Catholic Church from this important area of workers' lives; it also prevented religion from solving the "social question," understood here as the "worker question."[68] Leaders of the Catholic workers' associations therefore distanced their organizations from the Social Democratic unions and the Christian trade unions by seeking "gradual" improvements and "social reconciliation" by means of legislation, and they openly polemicized against the strike. Striking, Dr. Fleischer explained in a 1904 "social course," "endangers the natural duty of the worker to maintain his own and his family's existence" and "generates the most dreadful terrorism and endangers the personal freedom of coworkers under the threat of economic disadvantages."[69]

This approach converged with employer paternalism in another way: like the Catholic reform initiatives of the 1890s, it rested on an explicitly gendered ideological commitment to cultivating the private lives of male industrial workers—especially the "appropriate" gender order of the working-class family—as a means of controlling their public activity. The "cultivation [*Ausbildung*] of the worker," as one "social course" lecture pointedly argued, involved imbuing workers with the "virtues" associated with work and obedience: "diligence, loyalty, modesty, thrift, family values, and occupational pride [*Standesbewusstsein*]."[70] This meant sponsoring a range of family activities, from leisure activities to special courses designed, according to the league newspaper *Der Arbeiter,* to promote an "orderly domestic life" in working-class families, including a "sufficient, healthy, and friendly apartment," "a diligent, careful *Hausfrau,* who keeps everything clean and in order, who knows how to maintain a household, who understands how to keep domestic order and to prepare tasty meals with limited means," and "children who in their clothing and behavior manifest order and a good upbringing."[71] This moralizing project of cultural improvement was aimed at rooting out the "immorality and immoderateness" of the worker and the "immoral relations in the family," manifested in the growing "desire for pleasure" and consumer entertainments, the expansion of club life, and the rise of social democracy more generally.[72] In a 1904 speech before the Catholic workers' association in Neunkirchen, which detailed the league's methods for immunizing workers against the allure of pubs, commercial entertainments, and social democracy, Pastor Schütz of Wiebelskirchen made these connections between moral "education" and public life explicit: "One must give the worker the opportunity to find other kinds of enjoyments," he maintained, by providing lectures and appropriate reading materials to him. "If we want to have hope for success in the future," Schütz continued, "then it is necessary for Catholic workers to try to further cultivate and school themselves; in that way they will be in a position to support our cause in public life and be able to respond to the speech of our opponents."[73]

As the preceding quote suggests, Catholic labor leaders embraced the discursive strategies of bourgeois *Öffentlichkiet* and repeatedly called on "the public" to validate their own claims. The possibility that Catholic labor politics would provide a space for working-class self-assertion, however, was circumscribed by its reliance on gradualist proceduralism, official benevolence, and limited appeals to a well-meaning "public." In an exemplary formulation that implicitly demoted a "worker" interest in relation to a "public" interest, Catholic workers' association secretary Bartholomäus Koßmann, the leading activist of the associations in the region, explained to a 1909 miners' meeting in Neunkirchen that the demand for an eight-hour day in mining could not be realized simply by workers demanding it.

> The workers must also try to bring public opinion onto their side. I don't believe that the employer in question here would be so shameless—to use an unparliamentary expression—as to take away something from the workers in a way that seemed unjustified in the eyes of the public [*Öffentlichkeit*]. On the occasion of his visit to Altenwald, the former commerce minister Delbrück said: "If you have grievances, then you should direct them to me in Berlin, and I will help in so far as the complaint is justified." We want to take these words seriously and turn to Berlin if the mining officials here do not give us what we want. We would say, "Here now are our wishes; now keep your promise."[74]

This statement of strategy pointed to the avenue of official redress that favored the Catholic workers' associations and Christian trade unions over Social Democratic organizations, but it also indicated the limitations of appeals to the bourgeois public sphere that curbed the militancy of prewar Catholic labor in the Saar.

Despite the overall success of Catholic priests and lay activists in building workers' organizations with trade sections, they were not able to prevent the emergence of Christian trade unions in the Saar. Beginning in 1904, a full decade after the completion of similar efforts in the Ruhr, representatives of the Christian Miners' Union began attempts to organize

Saar miners.[75] Initially, *Gewerkverein* organizers managed to hold meetings only in the mining districts that lay outside of the Prussian Saar, but in the spring of 1904, the *Gewerkverein* leadership in the Ruhr sent trade union secretary Johann Effert and its general secretary Adam Stegerwald to the region.[76] The first meetings to attract members were held in the central mining towns of Altenkessel, Püttlingen, and Dudweiler on 1 May, and by June, the *Gewerkverein* had created eleven local branches with about six hundred members. By October, these figures had increased to nineteen locals with about twelve hundred members, and by the end of the year, after retaining Ludwig Buchheit as a local paid functionary, the *Gewerkverein* could claim four thousand Saar miners as members. In July 1905, the formation of a trade union cartel, which coordinated the activities of the locals and instituted courses for the training of union activists, completed the organization of the *Gewerkverein* in the region. The latter grew steadily during the prewar decade: by the beginning of 1911, the *Gewerkverein* had managed to organize 14,000 (or 20 percent) of the 50,802 miners employed by the Mining Office.

Like the Social Democrats, Christian union leaders in the Saar were able to build their movement by organizing through local public institutions, including the industrial courts, sickness funds, and the deliberative bodies in Saar mining. Already in 1905, Christian unionists were organizing for local *Ortskrankenkassen* and industrial court elections, not only to secure some control over the management of these institutions, but also to defeat Social Democratic candidates. They did this in cooperation with the Catholic workers' associations and other nonsocialist organizations, including the Protestant workers' associations. In 1909, the Christian unions led efforts to create the Social Council of Nationalist Employees' Organizations (Sozialer Ausschuss nationaler Arbeitnehmer-Organizationen), an avowedly antisocialist local formation, which included the Protestant and Catholic workers' associations and the liberal Hirsch-Duncker unions and was designed largely to forge coalitions capable of winning local elections to these bodies. By 1910, the Christian unions and their allies claimed to have "elected out" some 228 Social Democratic representatives of the *Ortskrankenkassen* in the Saar and to have taken over the sickness fund and the industrial court in Malstatt-Burbach, and they were entertaining hopes of taking control over the *Ortskrankenkassen* in Saarbrücken and St. Johann.[77]

Moreover, Christian union leaders exploited the status of Saar mining as a state-owned and state-operated industry and its role as one of the kaiser's "model institutions" that established mechanisms for employee representation in the "New Course" legislation of 1890–91. Accordingly, they not only benefited from the fact that nearly all Saar miners shared the same employer and relatively uniform wages and working conditions across the region; they also took advantage of the elective works councils (*Ausschüsse*) and the selection of security stewards (*Sicherheitsmänner*), institutions not available to workers in local private heavy industry and offering miners opportunities to influence decisions about wages, social provisions, and working conditions. As early as 1905, the *Gewerkverein* began to organize for elections to the works councils and for security stewards in Saar mining, and it gradually turned the councils into sites from which to criticize Mining Office policy during the prewar decade.[78] This activity was aided by the fact that crucial aspects of Mining Office policy were periodically aired as public matters, as its budget was subject to debate and approval in the Prussian Landtag—in sharp contrast to the conditions obtaining in local private heavy industry. This meant, as the Mining Office chairman complained, that working conditions and wages in Saar mining were made public and debated by opposition parties, including the Center.[79]

The relative success of the Christian trade unions in the Saar therefore owed as much to the specific constellation of discursive relations and institutions that determined access to the local public sphere as it did to the preponderance of Catholic workers in the region. Christian union leaders sought legitimacy by emphasizing their loyalty to the kaiser and the monarchical state and their commitment to moderate reform and proceduralist gradualism within the limits of "bourgeois society." This meant defending the interests of workers in the way other occupational associations represented the interests of their members, while affirming the basic parameters—and the attendant inequalities and constraints—of the capitalist economy.[80] Christian union leaders also formulated their aims in the gendered moralizing discourse of social Catholicism, which converged with the language of Protestant social reform after 1900. In an effort to define themselves as the decent and lawful alternative to socialism and to compete with the Catholic workers' associations on the terrain of moral instruction, Christian unions embraced a project that sought to immunize members against the contagion of dissolute forms of popular culture. They worked against the deleterious consequences of what Christian union leaders called *Klimbim,* meaningless pastimes and entertainments sponsored by the proliferating *Klimbimvereine* (i.e., carnival, bowling, sports, and conviviality clubs), as well as against the penchant for "trashy literature and pictures" and the immoderate enjoyment of alcohol, which constituted a "cancerous growth on the life of the Saar people."[81] Moreover, the focal point of Christian union rhetoric was the "family wage" (which defined workers as "family fathers") and its importance to the material and moral well-being of the working-class family. According to a *Gewerkverein* leader speaking in 1907, the absence of a "family wage" meant that workers' wives were forced to seek paid employment outside the home and to neglect their female obligation to "educate their children to be orderly and productive members of human society."[82] Finally, the Christian unions centered their approach on appeals to the public sphere as an authorizing agency, and in response to charges that they represented the "precursor to social democracy," they appealed to the arbitrating force of a reasonable *Öffentlichkeit*.[83]

This strategy was revealed in the weeklong Burbach steel strike of June 1906, the first mass industrial action of Saar workers since the winter of 1893.[84] Led by Leonhard Wernerus from Aachen, the secretary of the Christian Metalworkers' Union (Christlicher Metallarbeiterverband Deutschlands, hereafter CMV), the strike was aimed at improving the wages and conditions for employees at the Burbach steelworks and securing the right of workers to join their own union. Wernerus and the strikers adopted a strategy that centered on bringing "proof [of oppressive working conditions] before the public sphere" by means of an orderly presentation of grievances and attempting to send delegations, all of which were rebuffed, to meet with factory director Edmund Weisdorff.[85] This entailed explicit attempts to distance the "reasonable" demands of the CMV, put forward "only within the existing legal framework," from the "revolutionary" aims of social democracy and to embrace a universal language of "human rights" and citizenship for steelworkers over overtly confessional and class languages of protest, in an effort to challenge the paternalist forms of antiunion coercion at the steelworks.[86] In addition, CMV leaders formulated their demands in the gendered discourse of Christian labor, which directly invoked the steelworker's right to a "family wage" and accepted the bourgeois links between masculinity and "reasonable" public debate. By contrast with the dramatic crowd actions and self-mobilization of women during the miner's winter strike of 1893, the wives of Burbach steelworkers were called to two or three separate meetings, during which CMV leaders explained the importance of the union and urged the women to convince their husbands to join it.[87] Finally, Wernerus and other CMV leaders attempted to appeal to "the public" in ways that counterposed "citizens" (*Bürger*) to "workers" and implicitly accepted the ideological workings and hierarchies of bourgeois *Öffentlichkeit.* In his speech before a meeting of strikers on 8 June, Wernerus claimed, "The sympathies of the citizenry (Bürgschaft) are with the workers. We want to make sure, therefore, that the struggle is carried out by maintaining order and discipline." During the same meeting, Franz Wieber, representing the national organization, referred to the formation of a five-member commission to negotiate with management as a demonstration to "the public" (*Öffentlichkeit*) that "we seriously want peace and do not intend to damage the steelworks" with a long strike.[88]

Given these important areas of agreement with dominant definitions of moral and social order, it is not surprising that Christian union demands were allowed into the existing field of public discourse in ways that socialist claims were not. At the national level, the Christian unions gradually won the support of prominent bourgeois social reformers, imperial ministers, and the kaiser himself. Largely under the direction of Adam Stegerwald, the Christian unions established contacts with a wide range of Protestant Christian Social organizations (under the direction of Adolf Stöcker), Protestant and Catholic workers' associations, nonsocialist labor formations, and bourgeois social reformers associated with the Society for Social Reform, especially former Prussian commerce minister Hans Berlepsch. The most concrete public expression of this emerging community of sentiment was the congress of nonsocialist workers held in Frankfurt in October 1903. Organized by the Christian Socials but dominated by the Christian trade unions, the congress, which was followed by meetings in 1907 and 1913, was designed to impress on the "public" and government ministers the potential strength of the nonsocialist labor movement and to call for specific reforms, including the liberalization of the existing law of association, legal corporate status for trade unions, and the establishment of state-regulated works chambers composed of workers and employers. Well received by many politicians and their supporters in the bourgeois parties, including the Center, the National Liberal, and the left-liberal parties, the congress was praised by Arthur von Posadowsky-Wehner, imperial secretary of the interior, and the kaiser, who, via telegram, promised his support and protection for the assembled Christian workers as an important "dam against social democracy."[89]

Locally, the Christian unions were able, over time, to count on the support of nearly all of the leading newspapers. By 1910, all six major regional newspapers had opened up their pages to printed notices and news announcements of Christian trade unions' meetings and agendas.[90] More striking, most of the major newspapers in Saarbrücken and St. Johann—not only the Center-affiliated *Saarpost* and *St. Johann-Saarbrücker Volkszeitung,* but also the liberal *Saarbrücker Zeitung* and the left-liberal *St. Johanner Zeitung*—supported the Christian union-led strike at the Burbach steelwork in June 1906. Only the newspapers directly connected to heavy industry—especially the *Neue Saarbrücker Zeitung* and the "factory newspapers" in places like Malstatt-Burbach, Völklingen, and St. Ingbert—refused to tolerate the Christian trade unions.

Finally, the Mining Office's policy shift after 1905 proved most decisive to the prospects of Christian unionism in the Saar. The move to a position of neutrality in relation to the Christian trade unions brought an end to the paternalistic harassment of "Christian-national" miners. It meant that Christian unionists were allowed to hold meetings, publicize their aims, and even read their own newspapers, along with Center Party newspapers, in the mining dormitories and facilities.[91] Indeed, Saar mining officials not only allowed the *Gewerkverein* openly to seek members and publicize its demands; by 1913, they even recognized the Christian Miners' Union as a viable antisocialist formation, one that had the virtue of at least appearing to be a militant alternative to the socialist union. In a report assessing the potential costs and benefits of a company-sponsored union, Saar mining officials decided in favor of the *Gewerkverein* and against the formation of a company, or yellow, union, arguing that the former "manifests the necessary radicalism—even toward us—that will keep the masses from going over to social democracy and allow it to combat Social Democracy successfully," while "it is not so strong that it believes it is capable of doing without our well-meaning neutrality or of risking a strike without danger of self-destruction."[92]

This cynical strategy points up the limitations of the Christian labor movement and helps to explain its lack of success in the prewar Saar. Despite the fact that Christian union leaders mobilized nearly 4,000 out of the total of 4,480 employees of the Burbach steelworks during the June strike, their strategy of moderate interest representation ended in limited success, if not outright failure. The dispute was settled only with the intervention of two outside negotiators—the National Liberal Heinrich Boltz and the Protestant pastor Nold—in view of Weisdorff's refusal to negotiate directly with the union. In third-party negotiations, company officials only agreed to allow the striking workers to return to work if their jobs had not been taken by "faithful" nonstrikers, and they refused to rehire the fifty-six most active union organizers or to concede any general wage or workplace demands. Their only formal concession was to acknowledge the right of their employees to join a union, though they would not negotiate with any union officials not employed in the factory, nor would they "recognize any kind of obligation to negotiate with members of just any organization." They argued, "We will not, for example, hire Social Democrats or members of organizations that preach illegal breach of contract."[93] In this way, they were able to break the union completely. Subsequent attempts by CMV activists to organize steelworkers in nearby Brebach, Bischmisheim, and Völklingen and to compete against newly formed company unions at the Burbach, Völklingen, and Neunkirchen steelworks failed.

The Christian Miners' Union in the region was not very successful in developing an organization capable of addressing matters of economic or social inequality before the First World War. The *Gewerkverein* was never even able to secure for its members higher wages, which, by contrast with the early 1890s, lagged behind the wages of the Ruhr, Wurmrevier, and Upper Silesia during the entire prewar period. After 1905, the *Gewerkverein* expanded its organization and attempted to engage the Mining Office in debates over the reform of the miners' sickness and pension funds. Their petition, which they sent to the Mining Office, regional mining officials in Bonn, and the Prussian Landtag, met with very limited success, as did their subsequent efforts to enhance the efficacy of the mining works councils. Indeed, in the face of the economic downturn and deteriorating wage situation from 1910 to 1912, the five petitions, with accompanying memoranda, that the *Gewerkverein* sent to the Prussian commerce minister, the Prussian Landtag, and the local Mining Office secured only a promise of a future family pension fund.[94] While the Christian Miners' Union continued to win recruits during this period, the strike it threatened for December and January 1912–13 never materialized. After the mediation of National Liberal Party chairman Ernst Basserman, who won the seat for Saarbrücken in 1912, the dispute ended with only a minor concession, in the form of a promise of future wage increases from the Mining Office. Consequently, the *Gewerkverein*—opposed during these deliberations by the Catholic workers' associations, whose leaders argued that "public opinion" was not behind the strike—was hobbled by a mass exodus of its members.[95]

Public Life and Party Politics

The new civic associations, Catholic workers' associations, Christian trade unions, and Social Democratic organizations had more consequential and lasting effects on the paternalist organization of public life and party politics in the region. The efforts of the civic associations to open up the public domain of municipal affairs were carried much further by Social Democrats, Catholic priests and Christian labor leaders, and Center supporters, who began systematically to organize for council elections in the years after 1903. The consequences of this mobilization, which also included the Protestant workers' associations, were dramatically illustrated in a number of subsequent public controversies. In Neunkirchen, Catholic leaders successfully contested National Liberal influence over the town council during the municipal elections of 1907—which "for the first time brought party passions into the municipal council"—and ended the paternalist practice of candidate preselection with the approval of the director of the Stumm steelworks.[96] Subsequent elections were contested by all parties, and by 1910, the membership of the Neunkirchen municipal council included Bartholomäus Koßmann (an aforementioned leader of the Catholic workers' associations) and even a miner.[97] Perhaps more striking, in December 1911, a coalition of Center and reform-liberal members of the Saarbrücken municipal council prevailed over a coalition of industrial paternalists, including Director Heinrich Köhl of the Burbach steelworks, and granted a request submitted by the local SPD to use the main meeting hall of Saarbrücken, the Saalbau, for a party meeting in January 1912. After insisting that preventing the SPD from using the hall was no longer "appropriate to the times," the reform coalition effectively broke with the paternalist organization of municipal affairs and public life.[98]

The proliferation of working-class associations, the Catholic workers' associations, and the Christian trade unions also contributed to the momentum of the Catholic Center Party as it held to its policy on the social question and acted as spokesman for the interests of industrial workers in years after 1900. In 1903, the leaders of the Center Party in Saarbrücken cast themselves as the representatives of the workers against the employers, and their candidate Dr. Peter Muth supported the government's proposed tariffs against the right wing of the party's Reichstag fraction, called for the formation of trade unions out of existing workers' associations, and proposed unemployment support for workers. Against the backdrop of a doubling of the Social Democratic vote, Muth took a close second place to the National Liberal Boltz in the election. The "Hottentot" elections of 1907 were even more closely contested in the wake of the Burbach strike, and party activists with connections to *Saarpost* even pushed to have the Christian union secretary Wernerus selected as party candidate. The church and Center Party hierarchy rejected this effort, but their own candidate Dr. Strauss, who secured the support of Wernerus and campaigned on behalf of workers, lost by only 1,448 votes in the closest election in the district of Saarbrücken. More dramatically, during the Reichstag elections of 1912, the competing factions within the Center Party reached an agreement by which the Christian trade union secretary and miner Franz Sauermann from Duisburg would run as Center candidate in Saarbrücken and the Catholic workers' association secretary Koßmann would run in Ottweiler. Koßmann, the nonmilitant labor leader who was supported by the Catholic clergy, won the seat in Ottweiler and became the first "worker candidate" to represent the interests of a political party in the region.[99]

Finally, the rise of working-class associations and the social orientation of the Center Party, along with its electoral advances, pressured leaders of the National Liberal Party, long dominated by industrial paternalists and their allies, to consider a new "social orientation"—a change that partially reflected developments within the national organization, especially after 1905, when the Stresemann and Young Liberal left wing began to challenge the party's industrialist right wing.[100] After Stumm's death in 1901 and the dissolution of the long-standing "cartel," reform-minded liberals, including the Naumannite Protestant clergymen active in the mid-1890s, leaders of the Protestant workers' associations, organizers of the newly formed liberal Hirsch-Duncker trade unions, and editors of the *Saarbrücker Zeitung,* coalesced into a serious faction within the local National Liberal Party.[101] This faction was spearheaded after 1903 by the Saar Young Liberals, who established a regional organization and sought to challenge the industrialist domination of the local party by means of a reformist program of imperialism, social welfare, trade union rights, and a broad-based hostility toward political Catholicism.[102] Led by the lawyer Dr. Lanser and supported by *Oberrealschule* director and reform liberal Dr. Maurer, the Young Liberals challenged the party establishment to give up its intermittent and socially exclusive style and to bring workers "into practical collaboration in the service of the National Liberal Party" by recruiting working-class members and extending these efforts into some of the industrial towns and villages after 1904.[103] This prompted Boltz and the

National Liberals to expand the party's permanent organization (which comprised twenty-four separate electoral committees by 1906), hold regular meetings, and sponsor social events, to which even women were invited.[104] By the Reichstag election of 1907, party leaders were instructing functionaries to attend diligently to the publication of electoral notices in all local newspapers, the posting of placards on kiosks "on the streets" and in taverns and businesses, and the distribution of "*handbills to every house* in the districts for which your electoral association is responsible": "The cause will not be won by means of meetings, handbills, newspapers articles, etc. alone," party leaders insisted, "but will also involve above all the *necessary and most conscientious detailed labor* [*Kleinarbeit*], to which every member of your electoral association must enthusiastically dedicate himself."[105] But the most visible sign of the National Liberal "turn to the left" came in the wake of the Burbach steel strike,[106] during which the two main venues of reform liberalism, the *Saarbrücker Zeitung* and the *St. Johanner Zeitung,* supported the cause of steelworkers. Heinrich Boltz even helped to negotiate the end of the strike by calling on the director of the Burbach steelworks to acknowledge his employees' right to organize—a settlement that paved the way for Dr. Maurer to define the workers' right to organize in nonsocialist trade unions as a central campaign issue during the Reichstag elections of January 1907. This orientation was generally upheld by party leaders Ernst Bassermann, who visited the Saar during the elections in 1907, and Gustav Stresemann, who visited during the run-up to the Landtag elections in 1908, in order to hold the reformist and paternalist wings of the party together.[107]

Most striking was the change in National Liberal electoral rhetoric and practice between 1907 and 1912 and the new recognition of workers during campaigns—a novel attempt to incorporate workers into the disinterested consensus of the "liberal *Bürgertum.*" During the Reichstag elections of 1907 and 1912, Saar National Liberals began allowing some workers and even Christian trade union leaders to speak during general voter meetings.[108] During the 1908 Landtag election, which was the occasion for sharp clashes between the paternalist wing led by Boltz and the reform wing led by the Young Liberal candidate Maurer, National Liberal miners in Dudweiler held their own meeting in May in favor of the miner Pfordt as National Liberal candidate, in an impassioned debate over the social question and the worker's right to organize. The miner Müller directly challenged the electoral practices of Saar industrialists by calling for a miner to represent the local "liberal citizenry."

> On the economic terrain, workers are still not able to achieve equal rights in relation to a Weisdorff and a Röchling, so we want to achieve this even more on the political terrain. We are not mindless voters [*Stimmvieh*] but seek to be authors of political ideas and demand that our concerns are clearly and conclusively recognized. It is absolutely unacceptable that a Weisdorff lets Delegate Boltz give speeches about the worker's right to organize and then later throws unionized workers out in the street. That is political hypocrisy.[109]

Moreover, in response to the Center Party's strategy of putting forward the "worker candidates" Koßmann and Sauermann for the Reichstag elections of 1912, the Young Liberals even proposed a worker candidate, the blacksmith Pick who worked in the Fischbach colliery, for the Landtag election of 1913. Though these attempts failed, largely because most National Liberals could never accept a worker candidate, the candidacy and victory of Ernst Bassermann in Saarbrücken in 1912 and the direct appeals to workers were signs of dramatic changes within the local party in relation to the worker question. In the end, this turn to the left within the local National Liberal Party revealed most clearly the wider consequences of this period of working-class mobilization: the incipient collapse of paternalism, both as a labor regime and a framework for public life, in the Saar during the prewar decade.

CHAPTER 5

The New Managerial Rationality and the Racialization of Industrial Work

The fundamental reconstitution of Saar public life and the intensifying contestation over the meanings and boundaries of *Öffentlichkeit* initiated a process of ideological rearticulation that fundamentally transformed the managerial discourses of Saar employers after 1900. The structural interconnections between company paternalism and public authority meant that efforts to expand the conditions of *Öffentlichkeit* in the Saar necessarily, if indirectly, undermined the foundations of the paternalist factory regime. But the formation of working-class counterpublics and new spaces of opposition also enabled direct critiques of paternalist workplace organization after 1900. In response to Social Democratic, liberal reformist, and Christian-national challenges to paternalism, Saar employers began publicly and systematically to (re)define the nature of work and the relations between workers and employers in heavy industry. Abandoning or reaccenting the central coordinates of paternalist labor policy, they formulated a new rationality of workplace management, centered on the regulation of workers' bodies and the bioracial capacities of workers and their families, which sought to improve labor "performance" (*Leistung*) and overall productive efficiencies in the racial and economic "struggle" between nations and peoples. This new rationality was partially institutionalized and materialized in the architecture of Saar factories and in the labor relations and social policies of Saar heavy industry after 1900.

These new technocratic schemes of "rational" workplace management, which sought to discover and direct the allegedly natural laws immanent to human behavior and organizational forms, were part of a general American and European phenomenon within large-scale industry in the new era of mass production and global markets at the turn of the century.[1] They are usually grouped together under the label "scientific management" and associated with "Taylorism," or the ideas of Frederick Taylor, an American engineer whose theory of management based on the systematic study of work tasks and the creation of optimal workplace conditions and incentives for labor productivity were discussed widely among European engineers and industrialists.[2] But these schemes emerged from the writings of a broad array of engineers, employers and their representatives, social reformers, trade unionists, psychologists, sociologists, and physiologists across North America and Europe and from a wider set of concerns related to new divisions of labor and skill, the invention of new wage systems and incentives, the bureaucratic reorganization of managerial hierarchies, the introduction of new technologies and mechanized infrastructure, the reorganization of production processes, and the changing structures of the business enterprise as a "vertically" and "horizontally" integrated corporate structure after 1890.[3] In this context, new technically trained managers undertook efforts to organize the industrial workplace and the production process according to "impersonal" technical imperatives and to comprehend worker attitudes toward and physical capacities for work on the basis of scientific-technocratic considerations—efforts that were visible before the First World War, to very different degrees, in the U.S. electrical, chemical, and automobile industries; the British engineering and metallurgical industries; the French mining and automobile industries; and the German iron and steel, chemical, electrical, and automobile industries.

Nevertheless, the new scientific-technocratic orientation of industrial management took many forms and cannot be reduced to one general, ineluctable trend of "corporate" management in the modern industrial era. In Germany, it evolved out of two distinctive bodies of formalized knowledge: the "science of the factory" (*Fabrikbetriebslehre* or *Fabrikbetriebskunde*), or the production process; and the "science of work" (*Arbeitswissenschaft*), focused on individual worker attitudes and capacities. It also evolved in the context of specific political-ideological struggles over the organization of work in the large-scale factory. These struggles were shaped by the relatively greater concentration of German capital and the strength of German employer organizations and their newly established research institutes or "think tanks," in relation to organized labor; the higher levels of state involvement in social welfare institutions for industrial workers, especially after 1900; the increasing influence of reform organizations and the parties in the Reichstag (the Center, the left liberals, and the Young Liberal wing of the National Liberal Party), which could influence industrial legislation; and, above all, the presence of Europe's largest and most well-organized socialist party, the SPD, and its affiliated unions. In their immediate conflict after 1900 with the class languages of socialism and, to a lesser extent, Christian trade unionism, Saar employers developed their distinctive version of the new managerial rationality, which articulated technocratic ambitions with biopolitical racism.

Rationalized Management and the New Science of Work

The turn to a new rationality of workplace management, especially in the steel industry, eventuated from fundamental discursive shifts entangled with a series of changes, underway since the 1890s, to the forms of business organization, management structures and practices, procedures of cost accounting, and workplace technology related to matters of "efficiency" and profitability. It emerged out of a growing interest in (re)defining the economic and natural laws and properties governing labor relations and industrial capitalism and drew on the generalizing systematicity of scientific theory to produce a technocratic calculus of productive efficiency and workplace performance (*Leistung*) and scientific representations of the biophysical properties of labor and the capacities of individual workers, managers, and employers. But the timing of its emergence and consolidation as a discourse after 1900, its specific range of citation, and its veiled strategic claims suggest that this new managerial rationality should be understood neither as an unmediated or entailed outcome of technical imperatives and economic transformations nor as the inevitable application of scientific thought to the autonomous structures and relations of the industrial workplace. Rather, it is best understood as a response to competing discourses about workplace organization.

Saar and other German employers elaborated the scientific categories and assertions of this new rationality in ways that were designed specifically to debunk the definitions of work and the factory and the claims about the rights of workers and the most appropriate form of industrial organization emanating from the ranks of Social Democrats, bourgeois social reformers, and Christian unionists after the 1890s. Despite their limited organizational gains and their modest, proceduralist emphasis on securing the basic right to organize, Social Democrats launched the most systematic critique of paternalist work relations, which implicitly carried the utopian demand for the socialization of production and made social democracy the ongoing target—and haunting fear—of Saar industrialists. Bourgeois-liberal reformers, who found institutional support within the Young Liberals and the moderate wing of the local National Liberal Party, supported the formation of nonsocialist trade unions and "codetermination" at the workplace and helped to undermine paternalism partly from within, by invoking the paternalist aim of maintaining "peaceful relations between employer and employee" in order to justify its aims. Christian labor leaders, mobilizing Saar workers on the basis of a "class identity within a religious-ecclesiastical" framework, challenged Saar paternalism even more directly on its own terrain by reaccenting paternalist preoccupations with labor as a moral obligation, workplace harmony, appropriate familial order, and the male breadwinner norm in their efforts to empower workers as "citizens" with "human rights."[4] The accumulating pressure from workers and reformers in the 1890s and the trade union movements after 1903 turned industrial organization in the large-scale factory (*Grossbetrieb*) into an object of sustained political-economic struggle and focused national debates on large-scale productive relations and the general principles of workplace organization. In this context, Saar employers deemed the new scientific-technocratic rationality of factory management both attractive and imperative.

They were able to draw on an existing field of discourse about the organization of work in the large-scale industrial enterprise, expressed in the writings of a diverse array of academics, reformers, social investigators, engineers and technicians, and employers. This field had evolved into two discrete knowledges in Wilhelmine Germany: the "science of the factory" and the "science of work." The newly emergent science of the factory sought to "master and elaborate further upon the experiences gained heretofore through [managerial] praxis" and to articulate, "by means of scientific method," the principles and relations by which a factory was efficiently and productively organized.[5] It produced a body of "management literature," which began in the late 1860s as a series of isolated texts concerned with cost-accounting procedures but evolved into a dense field of knowledge after 1900, born largely by engineers and technical "experts" dedicated to defining and propagating general rules of industrial "organization." The latter were articulated in new technical journals—such as *Organisation. Fachblatt der leitenden Männer in Handel und Industrie,*

Werkstattstechnik. Zeitschrift für Anlage und Betrieb von Fabriken und für Herstellungsverfahren, and *System. Zeitschrift für moderne Geschäfts- und Betriebskunde*—and in the activities of trade organizations, especially the Association of German Engineers (Verein deutscher Ingeneure) and its own journals, *Zeitschrift des VDI* and *Technik und Wirtschaft.* These new publications and organizations aimed for and constructed an "interested public" of not only technical writers removed from the immediate domain of production but also factory directors, managers, and white-collar workers interested in the new science of workplace organization. Research related to specific factories as well as essays on cost accounting, bookkeeping, factory organization and communication, and general methods of labor management and payment schemes, including the time and motion studies and wage calculations associated with Taylorism, thereby entered into the journals of heavy industry and into the writings of publicists associated with the CVDI and other industrial organizations.[6] The new science of industrial management thus challenged antitheoretical paternalist understandings of work and factory organization, which focused on the unique economic talents and moral qualities of the employer and the workforce and the particular needs of and customary practices in the "private" individual factory. Indeed, the new science proposed generalizing theories about the factory and the production process—its properties and the laws governing its operation—to a wider, interested public.[7]

In the Saar, these "scientific" understandings of industrial organization informed several fundamental changes to the iron and steel firms and glass firms after the mid-1890s. First, they helped to displace the previous model of paternalist factory management (which involved the factory owner in direct relations with workers), by underscoring the potentialities of the limited liability firm and joint-stock company. The Burbach and Dillingen steelworks were already established as joint-stock companies—though the latter under the controlling influence of Stumm—before 1890, but new strategies of capital formation and organizational efficiency led to the transformation of nearly all leading Saar firms into limited liability companies—the Neunkirchen steelworks in 1903, the Halberg steelworks in 1908, and the Röchling steelworks, the Vopelius glassworks, the Reppert glassworks, and the Wentzel glassworks shortly thereafter.[8] This not only entailed the diminution of the "family spirit" allegedly governing earlier forms of paternalist business organization;[9] it also curbed the direct influence of the company owner, as the factory direction itself was divided into technical and financial sections and increasingly operated on the basis of regular rounds of managerial consultation. Second, the new scientific understandings informed the bureaucratic-technocratic reorganization of managerial structures of Saar industrial firms. Following a pattern manifested in industrial concerns throughout Germany and Europe, Saar firms evolved managerial hierarchies that not only departed from the paternalist concentration of control in the figure of the factory owner or director; they also weakened the authority of the foreman (*Werkmeister*) on the shop floor. In the place of more or less unmediated and command-style management practices, an expanded and attenuated factory hierarchy became the norm—a transformation manifested in the dramatic proliferation of technical personnel and white-collar clerical staff, whose numbers increased in Saar factories proportionately more than any other employees after 1900,[10] and of new offices and bureaus designed to perform a wide array of functions, including bookkeeping and cost accounting, sales and marketing, the procurement and allocation of raw materials and equipment, the calculation of wages and payments systems, the policing of factory grounds and personnel, the administration of social welfare programs, and the organization of company publicity and public relations.[11] In this context, the "*Werkmeister* system," in which a foreman directed each workshop or section in a factory and distributed tasks among his workers and set wage levels according to his experience with the ultimate approval of the factory director, increasingly gave way to rational-bureaucratic systems of wage calculation directed by engineers and technical and financial experts. This process was enabled by the systematic reorganization of procedures for setting wages and by the introduction of new practices and technologies, including filing systems and work cards, replacing the older, workbook system. This allowed direct comparison between old and new work tasks and the abstract-rational, rather than customary or experiential, determination of workers' wages and tasks.[12] The determination of job categories, work tasks, and wages were thereby subjected to a technocratic calculus of rational measurement and to systemic concerns about financial management under the auspices of an internal factory bureaucracy composed of technicians and clerks.[13]

Third, the new science of the factory encouraged the introduction of new payment schemes, especially in the iron and steel industry, which increasingly subjected production workers to the anonymous pressures of bureaucratized decision making and wage incentive systems—technologies that privileged impersonal, individualizing, and internalized norms of worker self-discipline over the direct commands of factory management. This was most obviously reflected in the growing prevalence of the "accord system," or piece rates, and the "premium wage." The accord system fixed the amount of a worker's wage according to quantity of output produced; each individual worker was then paid differentially according to a particular set calculus or hierarchy of tasks. After having displaced the hourly wage for most categories of workers by the 1890s, the accord system was enhanced by the "premium wage" system, which came into comprehensive use around the beginning of the twentieth century. The latter comprised a minimum base wage and a variable component hitched to the level of labor productivity—a form of wage incentive that encouraged workers to intensify their individual output.[14] These forms of payment were devised in the context of the complex restructuring of pay and skill levels in the metallurgical industries that took place after the 1890s. As Thomas Welskopp demonstrates, pay levels among workers underwent processes of both homogenization and differentiation: overall differences between the most skilled and least skilled workers narrowed, but the gradations of difference between pay levels and within broad job categories proliferated.[15] This process was directly related to the fundamental shift in the overall composition of the labor force in the Saar steel industry during the 1890s, when the combination of artisanal workers (e.g., puddlers and smiths) and unskilled workers gave way to a combination of skilled workers, unskilled or casual laborers, and a majority of new, semiskilled workers. The overall dilution of skill was registered in the proliferation of job titles among Saar wage earners—the Röchling steelworks reported over 150 categories of work in its various plants (i.e., the blast furnaces, the steelworks, the rolling mills, and the metalworking and ironworking shops) in 1909—and their apparent irrelevance to actual tasks performed.[16]

Finally, these changes to factory hierarchy and management were associated with the introduction of new machinery, workplace technologies, and production processes (particularly in the iron and steel, metalworking, and glass industries), which also undermined the "personal" dimensions and direct managerial relations of the paternalist labor regime. The manufacturing "system" of Saar iron and steel firms entered into a process of qualitative transformation in which the system of centralized, partially mechanized, and serial production, which prevailed since the 1880s, gave way to a system of decentralized integrated production during the prewar decade. The latter, as Thomas Welskopp demonstrates in excellent detail for steelworks elsewhere in Germany, was characterized by increased "flexibility of production levels, hierarchization of manufacturing processes, and simultaneity of production flows."[17] It was made possible by the introduction of new machinery and technologies (including cranes, lifts, ramps, and conveyors, which were increasingly powered by electricity) and the general spatial and architectonic reconfiguration of Saar steel factories after the turn of the century. As a comparison of the architectural designs of the Burbach steelworks of 1881 and 1906 illustrates, the rationalized industrial concern came into being with the vast expansion of industrial plants and the construction of multiple rolling mills, steelworks, workshops, and administrative-technical offices (including a research laboratory), connected by new conveyors and transport systems that traversed the entire complex of the factory grounds.[18] Similarly, the Röchling concerns expanded into a vast campus of independent factory plants and sections linked by new systems of transport after 1900, including a workshop for motorized lathe, drilling, and planing machines; a construction and carpentry section; an ironworking shop; an equipment storeroom; a chemical laboratory; and an electric power plant.[19] This architecture was the product of the new imperatives associated with decentralized and simultaneous flow production, which brought the qualitative transformation of the internal workings and processes of the paternalist factory and rendered the modest, localized, and centralized architecture of the paternalist enterprise a thing of the past.[20]

The second and related field that contributed to the new managerial rationality was the new science of work, which involved the systematic study of the relationship between human beings and the mechanics of industrial labor and developed as a European-wide phenomenon in the 1890s. In Germany, it emerged as a cross-disciplinary field of research, evolving out of the natural and social sciences, such as biology, medicine, psychiatry, physiology, and sociology. Its most preeminent practitioners were the physiologists Otto Fischer, Christian Wilhelm Braune, Nathan Zuntz, and Ludwig Max Rubner; the psychologists Emil Kraepelin and Hugo Münsterberg; and reformers and sociologists associated with the Verein für Sozialpolitik, including Max Weber, Alfred Weber, Gustav Schmoller, Heinrich Herkner, and Marie Bernays. In their very different ways, these academics and investigators examined mechanical operations in the workplace, the muscles and movements of the human body, the "psychic processes" of the human mind, and the attitudes of workers, subjecting all these factors to extensive laboratory investigation, photographic studies, and quantitative and qualitative social-scientific measurement.[21]

The result of these studies, as Anson Rabinbach makes clear, was a powerful and pervasive scientific representation of labor as the "expenditure and deployment of energy," an objectively quantifiable physical force expressed variously in nature, the machine, and the human body.[22] By contrast with earlier discourses about labor, which stressed its moral and spiritual dimensions, work was now universalized as a category of energy and framed within a wider scientific calculus that emphasized the physical economy of labor power (*Arbeitskraft*). Thus the science of work shifted attention from the moral properties of labor and the laborer to the physical tasks of work and the physiological and psychological capacities of the laborer's body. It became preoccupied with the human body's psychophysical capacities and limits—the biological foundations of exertion and fatigue—as experienced by the laborer on the shop floor.[23]

This new scientific turn in discourses about the factory and industrial work structured a wide range of politically diverse knowledge about the labor process, but its vocabulary and conceptual apparatus proved especially attractive to a group of economists and academically trained publicists associated with the organized public relations efforts of German heavy industry after 1900. Conservative intellectuals eagerly embraced the putatively nonpolitical and value-free claims of science in an effort to challenge what they viewed as the "moralistic" assumptions and reformist bias of the leading professors of German political economy (i.e., the so-called *Kathedersozialisten*, or "socialists of the chair"), the German labor movement, and state-sponsored *Sozialpolitik.* From the ranks of academia, the two critical figures in this industry-friendly scientific research were the University of Breslau professor Julius Wolf, who began publishing scholarly studies and newspaper articles critical of social reform in the 1890s and expanded these efforts with the creation of the *Journal for Social Science* (*Zeitschrift für Sozialwissenschaft,* hereafter *ZfS*), which he founded in 1898; and the University of Frankfurt economist Ludwig Pohle, who received the financial support of heavy industry and published a number of theoretical works on political economy and who took over Wolf's editorship of the *ZfS* in 1910.[24] The second most important center of this activity formed around University of Rostock economist Richard Ehrenberg, whose close ties to heavy industry went back to the 1890s, when he was director of the Commercial Collegium in Altona. Ehrenberg subsequently linked his academic ambitions with the interests of German employers, through his research projects on the Krupp steelworks in Essen and the Siemens electrical firm in Berlin; his journal, *Thünen-Archiv,* begun in 1905; and his research center, the Institute for Precise Economic Research, created in 1909 and financed by the CVDI.[25]

Finally, the academically trained industry spokesman and Pan-German Alexander Tille, who succeeded Stumm as the managing director of the main industrial organizations of the Saar in 1903, created the third important site of publicistic activity dedicated to the interests of heavy industry. Shortly after his arrival, Tille reorganized the editorial offices of the Saar industry newspaper and orchestrated its redesign, established a local business library and archive, and launched three important publication series designed to respond to the claims of the German labor movements, social reformers, and state social policy. These included *Südwestdeutsche Wirtschaftsfragen, Südwestdeutschen Flugschriften,* and *Sozialwirtschaftliche Zeitfragen,* which published book-length essays dedicated solely to defining the properties and laws and relations governing the industrial workplace and the global capitalist economy and combating socialism in "all its forms." Tille, who was a member of the Pan-German League and the Free Conservative Party, was a tireless publicist on behalf of Saar and German heavy industry. He served as assistant to Axel Bueck, chairman of the CVDI, from 1901 to 1903; and after taking up his position in the Saar, he remained a member of the CVDI's executive committee, where he subsequently pushed for the creation of private "academies for the schooling of agitators and teachers," for the purpose of training speakers to propagandize on behalf of heavy industry.[26] As the leading representative of Saar industrial organizations, Tille spoke for Saar industrialists generally and was instrumental in articulating the new ideologies of work and factory order in the region.[27]

In their different ways, this new cohort of conservative academics and theorists attempted to ground German political economy and workplace management more firmly and "objectively" in the methods and "discoveries of the natural sciences."[28] In their evolving corpus of "objective" economic theory, they defined work both as a product of energy (*Kraft*) or drive (*Trieb*) and as a process involving the transformation of nature into goods of value according to the dictates of a "productive goal" and "the application of technical processes for the achievement of that goal."[29] In this process, according to Wolf, the industrial economy had produced three different forms of human labor: "creative" (entrepreneurial), "dispositional" (administrative and managerial), and "executive" (manual) labor. Each form of productive activity was based on varying levels of "physical, deliberative, intellectual, imaginative, and technical talent," though the clearest distinctions were between "mental" and "manual" labor.[30] This opposition between the mental and manual dimensions of work, linked to the emphasis on the centrality of entrepreneurial or creative intention, explicitly served to refute reformist and union claims about the importance of wage labor and to elaborate a new vision of labor hierarchy in the modern factory. Thus Ehrenberg argued that the growing size and complexity of the "modern industrial enterprise," which was measured in terms of its division of labor and the application of machines and technology to the manufacturing process, rendered managerial skills the decisive motor of modern industry. Such "intellectual labor" was far more important than manual labor, which played only a minor and increasingly insignificant role in production.[31] Tille also privileged the "mental" labor of the employer, the "bearer" of the "productive economy," over the manual labor of the worker (*Handkraft*), since the latter is "a muscle power like animal muscle power and might be capable of nothing other than the latter, namely, the ability to carry out movements."[32] In this scheme, the employer appeared as the truly "creative force" in society, since only the presence of entrepreneurial intention created work in the first place: other "operations of manual energy," such as play, sport, exercise, or "horsing around" (*Unfug*), lacked direction and were not, strictly speaking, work. "The 'work' of the factory laborer," Tille maintained, "is not at all produced by the bearer of labor power [*Handkraft*] . . . but by [the employer] who prescribes what the worker has to do."[33] In this process of refuting the claims of social reformers, trade unionists, and Social Democrats in the language of "science," the ideologists for German heavy industry universalized work, the worker, and the employer, now understood in terms of a general category of producer (*Unternehmer*).

Both the scientific conception of work as a biophysical property and the categorization of work according to the varying and hierarchical functions of workers, managers, and employers were increasingly visible in the language and practice of German employers in heavy industry. In the Saar, especially after the arrival of Tille, this change was particularly evident in the local industry press, in industry-sponsored publications, and in the general reorientation of employer politics. As early as 1901, even before Tille's arrival, articles in the weekly *Neues Saarbrücker Gewerbeblatt* about the development of factory industry in Germany linked the fate of German business to the development of machine technology and the ability to attract high-quality *Arbeitskräfte* (labor power) or *Handkräfte* (hand power), terms routinely invoked in reports of the Saar Chamber of Commerce, company brochures, and Saar industry press articles after 1905.[34] Workers were now increasingly defined in terms of physical properties, production factors measured and evaluated according to their biological and mental "capacities" for labor, which resulted in assumptions about the efficiency and performance of workers and machines as interchangeable components in the manufacturing process. A 1910 report from the Saar Chamber of Commerce, for example, described the benefits of new machinery in terms of the reduction or "saving of *Handkraft,*" and a company brochure from 1905 assessed the "performance capacity" (*Leistungsfähigkeit*) of coal-washing machinery in a steel factory in the same way that it discussed the labor of individual workers.[35] In contrast to the primarily moralizing reference to the intimate relations within a company "family" figured in paternalist discourse, this new economic and biological calculus recast the workplace as the site at which machines and men came under the direction of the employer, or the bearer of "intellectual power," and were harnessed to a "productivist will" and toward a "productivist goal."[36] The critical function of the employer now involved assessing the physical and mental qualities of the labor force from the "productive-economic" (*kraftwirtschaftlich*) standpoint.

> He must fix the wage high enough to attract the laborers [*Handkräfte*] that he needs. He must try to attract the best labor power that he can maintain at the given level of wages. He must find the means to place each [worker] in the position in which the laborer's capabilities are best exploited; he must encourage the worker to perform at the highest level he is capable of; and he must know how to preserve the worker in order to maintain his labor power as long as possible.[37]

The new workplace calculus meant paying attention to the abilities, skills, and physical limits of different workers performing varying tasks in the factory; and it meant restructuring the workplace by finding the appropriate fit between workers and their jobs, by instituting a sufficient number of work breaks, and by installing hygienic facilities to maintain the physical health of workers.[38] Labor policy in Saar heavy industry, in other words, was reconfigured around the notion of the laborer as an agent of physical force, whose physiological capacities, general conditions of health, and functional relationship to the tasks before him or her were the principal concerns of the employer.

Perhaps the most revealing expression of this managerial rationality was the set of factory work rules published by Saar industry in 1913. In a striking contrast with the work rules and "disciplinary regulations" favored by Stumm and other paternalist employers from the 1870s to the 1890s, with their excessive concern with workers' morals and political activities and their assumption of a tutelary posture for employers, the new "model work rules of Saar industry" of 1913 stressed the contractual aspects of employer-employee relations, omitted any mention of the outside activities of workers, and limited the direct authority of the employer to the workplace. The only reference to "political, religious, and trade union activities" came in clause 75, which prohibited the collection of money, the distribution of handbills, and any discussions, meetings, or assemblies related to workers' organizations within the factory workplace or on company property—a clear retreat from earlier paternalistic claims on the right to control workers' trade union and political activity in the community and at home.[39] The new rules also defined the wage relationship between employer and employee in strictly functional and legalistic terms: the process of managerial instruction and consultation and the procedures for addressing workers complaints were described in terms of the hierarchy of competencies and "instances" within the factory. In a turn away from the "personal" contacts between factory owner and worker, realized most visibly in Stumm's weekly office hours, the new work rules elaborated a "process for lodging complaints" that followed a strict managerial hierarchy: "It is not permissible," the model statutes intoned, "to come without invitation to the top factory management."[40] In addition, the new model rules assimilated the terms and language of "rational-scientific" management: they rested on an understanding of work relations, wage scales, and company benefits as products of the interaction between economic laws and the function, use, and "performance" of *Handkraft*. Section 91 described the "contractual obligations [*Leistungen*] of the employer."

> For the faithful and precise fulfillment of his occupational duties in the factory, the wage earner is provided a wage, which differs according to the nature of the task and is set by scale according to level of performance [*Leistung*]. The wages of individual plants are determined conventionally but will change from time to time as the relationship between supply and demand affects a specific kind of labor [*Handkraft*]. They are valid as long as they have not been expressly changed by the employer [*Unternehmer*]. . . . Some come in the form

> of daily wages, some in piece rates, some as daily wages with performance bonuses. In addition to these, there are bonuses for punctuality, bonuses for consistency [*Regelmäßigkeit*], bonuses for matter-of-fact efficiency [*Nüchternheitsprämien*], according to the determination of the respective postings, as well as bonuses for expertise [*Sachprämien*] through participation in welfare programs. The kind of wage a worker receives will be specifically negotiated with each individual worker.[41]

Finally, as a "model" set of guidelines devised in anticipation of upcoming legislation related to work rules throughout Germany, the new work rules embraced the generalizing dimensions of the new managerial rationality. Tille, listed as "editor" (*Bearbeiter*), had been charged with the task of publishing the new rules as a model for all Saar firms, by the leading industrial organizations of the Saar—the VWGWISI; the SGVDESI; and the Employers' League of Saar Industry (Arbeitgeberverband der Saarindustrie), formed in 1904—and their leaders, including Louis Röchling, Edmund Weisdorff, Richard Vopelius, Dr. Hallwachs of the Wendel glass factory, factory director Müller of the Stumm concerns, and factory director Saeftel.[42] The rules were published in the series *Südwestdeutsche Flugschriften* and thus designed to appeal to a wider "public" beyond the individual factory or even the Saar region. In this way, the new model work rules were the collective product of consultations among Saar industrialists, as opposed to the private individualized concern of each factory owner; and they were premised on the scientific-technocratic claim to describe general properties and functional imperatives that allegedly governed relations between workers and employers in all business enterprises. No longer reducible to the personal qualities of the individual employer, practical considerations, and local particularities, the organization of work in the large-scale factory was now understood in terms of generally valid and scientifically derived laws related to the mental and physical capacities of employers, managers, and workers.

Biology, Race, and Industrial Labor

The growing interest in the rational workplace management and technocratic managerial strategies articulated with newly emergent bioracial theories and assumptions about work and production in heavy industry in the decade after the turn of the century. This process coincided with the formation and growing popularity of new fields of knowledge, such as eugenics and racial hygiene, concerned with the rational management of population size and strength and the healthy reproduction of the German nation or "race." It was part of a broader tendency, contested by confessional, liberal, and Social Democratic reform discourses, toward the "biologization" of the social, which pervaded the political culture of the nationalist middle classes in late Wilhelmine Germany and reoriented a wide range of disciplinary knowledges and political ideologies around questions of race, production, and reproduction.[43] In the Saar, the turn to "rational" management brought new preoccupations with the physical properties of the individual worker, the bioracial composition and activities of the worker's family, and the biopolitical relationships between the worker, the working-class family, and the wider health and well-being of the German "national body" (*Volkskörper*).

The most systematic connections between work, biology, and race were made in the evolving discourse of conservative political economists and employers in the decade before 1914. In attempts to ground economy theory in the "spirit of modern science" and especially in biology, Adolf Weber and the aforementioned group of right-wing ideologists—that is, Wolf, Pohle, and Tille—helped develop hereditarian and biological understandings of labor in which work became more than the expression of energy and physiological factors per se; it was also the product of innate "capacity" and "talent" (*Begabung*). According to Wolf, the capacity for creative, organizational, and manual forms of labor lay in the varying and inherited physical and intellectual talents of the individual, which were distributed unevenly throughout the population: the masses were mainly composed of the less "gifted" (i.e., manual laborers), while the "creatively gifted" (i.e., employers and entrepreneurs)—those with the most "value"—constituted an increasingly smaller percentage of the population.[44] According to the Pan-German Tille, this "natural economic diversity" of humankind meant that individual workers were capable of varying degrees of job "performance" (*Leistung*), and the quality of each individual's performance in the realm of productive labor became the standard measure of the individual's innate capabilities and "competence," as well as the primary reason for rewarding some workers with higher wages and advancement within the factory and maintaining others at low wage levels.[45]

In this scheme, all aspects of productive and social life were ordered around what were deemed the essential biological—and therefore gender and racial—hierarchies of nature. According to Tille, in the "productive (or profit) economy," competition among men resulted in a social hierarchy determined on the basis of inherited abilities and job performance. The top of this hierarchy, far above the "broader masses," was occupied by the "social aristocracy" of biological endowment and achievement. In the productive economy, the figure of the employer was the true "social aristocrat," the "man who under his own steam and his own responsibility and liability confronted the dizzying array of economic conditions and wrestled from them a profit or a loss."[46] Moreover, in the "savings economy" of the household, argued Tille, the "natural meritocracy of mental and physical attributes arranges itself . . . in a way that subordinates the woman to the will of the man and the children to the will of both parents."[47] In this view, the husband-father was best equipped to take on the risks and duties associated with ensuring the economic survival of the family unit, and the wife-mother was responsible for the continued biological reproduction of the family.

Finally, this discourse extended these biological, gendered, and labor hierarchies to the spectrum of "races," in a world allegedly founded on the interarticulations between race and capitalism. According to Julius Wolf, the "structures of the world economy" and the economic struggle among the nations were consequences of "racial characteristics," the "product of the blood mixture out of which nations had been created."[48] The Pan-German Tille, more than any other right-wing intellectual during this period, similarly situated the dynamics of modern factory production and industrial labor within the context of the international "struggle for existence" between the "races."[49] In this ideological framework, the rational, masculine, and advanced intellectual qualities of the European "master races" (*Herrenvölker*) were counterposed with the lesser developed, "feminine," and animal-like qualities of non-European peoples. In the realm of work and "productive efficiency" (*Leistungsfähigkeit*), therefore, the "indo-Germanic race" and the white races in general were deemed racially superior to the "Mongol," the "Negro," the "Australoid," and the "Indian" races. This superiority was measured in terms of the higher levels of "entrepreneurial spirit" and "commercial sense," the advanced intellectual and organizational talents, the enhanced disposition to save and economize, and the more pronounced technical skills, "nimbleness," and physical dexterity of the Europeans. By contrast, the Mongols (i.e., the Chinese and Japanese) were thought to possess only a "large measure of imitative talent" and "little sense for innovation in mechanical labor." The Chinese in particular were thought to lack the qualities and discipline for savings and the accumulation of wealth, and it was expected that their excessive "desire to play games of chance" and penchant for "sexual excess" and for dissolute behaviors, such as smoking and opium use, would permanently keep them in poverty. The "Negro" and the "Australoid" were deemed closer to "animals, who eat as much as possible and work as little as possible."[50]

Despite their racist condescension, these judgments and schemes of classification were expressed in a wider discourse of political economy that rested on fear of global economic competition. This was perhaps most visible in representations of certain East Asian peoples, such as the Chinese, who were portrayed as virtual automatons in their practically inhuman capacity for unending hard labor, and the Japanese, who were viewed as clever imitators.[51] But this fear also generally informed expansionist demands for access to world markets and the acquisition of colonial possessions in places like Africa, especially among the ranks of German heavy industry, as the critical means by which to sustain the necessary conditions of capital accumulation. In this sense, the racism of employers articulated fears about the productivity of individual German firms and industries and demands for imperialist expansion and colonial dispossession in the increasingly competitive environment of an expanding and globalizing industrial capitalism.

In this context, German industrialists framed their connections between race, gender, and work in a political economy increasingly grounded in social Darwinian conceptions of evolution and race struggle in the decade before 1914. This trend was evident in the 1900 Krupp essay competition on racial hygiene, in the publicistic activities of Krupp director and press magnate Alfred Hugenberg, and in the growing currency of racial-nationalist terms and appeals in the press organs of German heavy industry, where references to "racial pride" and "race struggles" appeared with growing frequency during the final prewar years.[52] But it was most visible in the writings of Saar industry spokesman Tille. Perhaps more systematically than others, Tille imagined economic life as the central realm of human activity, grounded in nature and operating in accordance with the biological struggle over the "means of existence" and the law of "natural selection." Each individual man, according to Tille,

> is conceived and born like the animal and can only prosper and maintain himself when he succeeds in securing the necessary means of existence in the long run. This happens in the productive economy, in

> which he performs economically profitable work for which there is a need. This work might be pleasant or unpleasant; he must do it just the same, if he wants to fulfill his first existential obligation and make it through life.[53]

The workings of the industrial economy therefore mirrored conditions in the animal world. Economic life, Tille claimed, was "unmerciful," "without sympathy, hard and cruel." It "recognized only forces, not opinions," and it knew "nothing of virtues and earnings." It knew only the innate "aptitude" and "lack of aptitude" of the individual.[54]

Accordingly, work and productive life became the critical terrain of evolutionary change and the development of racial fitness. In the realm of productive labor and economic competition, individuals could realize their inherited capabilities. The workplace, according to this view, was an arena of ceaseless competition between differently endowed individuals capable of varying levels of job performance (*Leistung*). Those capable of higher levels of performance—employers and managers—represented the most racially "fit," while lower-level clerks and wage earners filled out the ranks of the less "fit." Moreover, the industrial economy provided the "fit with considerably better conditions and encourages their propagation, since it is unable to extend its generative power to the unfit."[55] The result of this interaction between biology and unfettered productive relations would be the privileging of the social aristocrat of productive labor and the gradual improvement of the German *Volkskörper*.[56]

Yet the "productive economy" interacted with the "household savings economy" in ways critical to the operation of natural biological laws. First, the gender order of the household economy ensured that women would be available to give birth to and nurture the young; they would be able to function as the caretakers of the future generation. Second, "the orderly, regular life of the husband in the house always had a positive effect on the concentration of his powers [*Kräfte*] and the increase in his labor efficiency [*Leistungsfähigkeit*]." Finally, the combination of risks and responsibilities associated with setting up the household—the existential matters related to family support and maintenance—acted as a constant "inducement to greater effort." As such, they sharpened the skills and drive of the husband-father and better equipped him for the wider struggle in the productive economy.[57]

This turn toward hereditarian and evolutionary notions of race struggle in the discourse of German employers was linked to the shifting relations between unions, employers, and the state, as well as the changing political debate over state social policy after 1900. During the decade before 1914, leading sectors of German heavy industry, especially the large extractive and capital goods sectors represented by the CVDI, grew increasingly restless with the expanding influence of the labor movement (both socialist and Christian), the impressive success of the Social Democratic Party in the 1903 elections, and the professed reluctance of the imperial government to introduce repressive labor legislation in the Reichstag. Above all, German heavy industrialists focused their attacks on the expanding scope of state regulation of the factory workplace and *Sozialpolitik.* The Bismarckian social insurance schemes (i.e., the sickness, accident, and old age and disability programs) and the mechanisms established for workplace arbitration (i.e., works councils, industrial courts, and the proposed chambers of work) were increasingly deemed overly costly, invasive, and burdensome to employers. The sickness funds, work councils, and industrial courts in particular had been, according to this critique, abused by workers, who took advantage of the insurance benefits in illegal ways, and by the socialist and Christian union organizers, who had contested elections to their governing boards, turned them into mechanisms of union agitation, and used them to press claims against employers. Moreover, the CVDI criticized the way in which social insurance schemes had encouraged workers increasingly to claim benefits as their legal right, rather than to view them as paternalist offerings from a benevolent state.[58]

Once among the leading advocates of state social welfare, Saar employers (along with many other leaders of German heavy industry in the CVDI) now criticized its reformist foundations, claiming that it privileged social and ethical factors over economic, biological, and racial considerations. In the Saar, attempts by Christian and socialist unions to establish works councils, collective wage agreements, and unemployment funds for workers and to contest elections to local sickness insurance boards as early as 1903 were dismissed as "economic moralism," a dangerous attempt to impose ethical criteria or an artificial egalitarian vision on an industrial workplace shaped by natural economic forces and biological and racial laws of inequality.[59] In general, Tille argued, reform measures tended to extend benefits to the least productive and least valued members of society, those lesser humans who, "through gradual dissipation" and degeneration, had begun their "descent to the level of swine" and would soon "die out."[60] In addition, opponents argued that excessive social insurance benefits for all industrial workers would end in the "incapacitation" of the wage-earning classes, since such material guarantees would encourage "thoughtless living, indebtedness, a lack of concern for the future, and the notion that savings were not necessary." State welfare provision, according to this critique, undermined the skills necessary for workers to manage their household economies and to attain economic independence.[61] In this view, state welfare, by extending protection and benefits to the biologically "unfit" and by undermining the capabilities of the "fit," violated the law of natural selection, inhibited the development of the social aristocracy of work, and threatened the racial health of the nation.[62] This view did not represent a wholesale rejection of "morals"—Saar employers were still accusing striking and Social Democratic workers of being "immoral," and Tille was attempting to define a new "morality" of race—but its racial orientation did lay stress on the proper management and cultivation of physical bodies.[63]

These concerns about productivity and racial fitness were also central to changing employer attitudes toward company social provision in the decade before 1914. In the Saar, where generous company welfare and its moral benefits were once celebrated by local industrialists like Stumm, employers now routinely warned of the excesses of factory welfare and its potential danger to productivity. As in the case of *Sozialpolitik,* they complained of the ways in which trade unions and social reformers had turned the provision of company welfare into an employer's "obligation." Indeed, after 1900, organized Catholic workers in the Saar showed much less gratitude for company "benefits." They came to regard factory welfare as a negotiable condition of employment and thus subverted the function of moral tutelage that formed the basis of such welfare schemes.[64] As a result, local employers claimed, the excessive provision of company welfare made unavoidable the distribution of "unearned benefits" to employees.[65]

Rather than reject state social policy and company welfare outright, however, Saar employers advocated calculated measures of state and employer intervention designed to ensure the proper operation of the mechanisms of racial selection. This meant securing the "dynamic natural meritocracy of mental and physical attributes" both in the household economy, by subordinating women to men, and in the productive economy, by subordinating "the intellectually weaker to the intellectually stronger, the physically weaker to the will of the physically stronger, and the physically strong also to the will of the mentally strong."[66] Thus, if large numbers of potentially useful workers manifest an "incapacity for savings and efficiency, an irresponsibility and thoughtlessness regarding their health, bodily capacities, and capabilities," the state should intervene with more limited measures designed to promote self-help, thrift, and physical health. In this way, state-sponsored and mandatory insurance and sickness schemes would "sustain the labor power of wage earners [*Handkräfte*] for a longer period of time" and would not be concerned with the excess "comfort" and luxury of workers. In addition, labor protection laws for women and old age and disability insurance schemes for injured men would reduce the concern of the "working sex" (men) for the "work-weary sex" (women) and thereby free women to provide better care for children. They thus ensured the necessary means for the healthy development of the younger generation. All of these forms of intervention, according to Saar employers, should be geared toward generating "healthier and fitter" workers and optimal levels of economic productivity, which in turn would contribute to the "health of the nation."[67]

In addition, in 1908, Saar employers began to call explicitly for the "reformulation of the function of factory welfare institutions" and to reimagine company social policy as a means of fostering racial development. Rejecting the notion that the employer had any obligation to provide benefits and welfare for workers, they now viewed company social provision as part of the wage calculus—that is, it was redefined as "compensation in return for superior performance"—and framed it within the rhetoric of racial economy. Indeed, as early as 1903, the recognition that the workplace was defined by biological aptitude had led Saar employers to insist that the modern industrial workforce was composed of an upper stratum of "skilled industrial workers" and a lower stratum of unskilled casual laborers (*Handlanger*). They went to great pains to identify this "social aristocracy" of skill in the factory by means of surveys and quantitative measures: the "skilled" and "capable" industrial worker, a product of modern technological-industrial change, was defined on the basis of his length of training (usually one or more years), wage level (twelve hundred to sixteen hundred marks per year), full membership status in the company *Knappschaft,* mental and physical qualities (e.g., "endurance, attentiveness, conscientiousness, dexterity, deftness, presence of mind, and consideration"), patterns of socialization, and alleged self-image as a disciplined and proud worker.[68] Accordingly, "special advantages," including housing, loans, and subsidies, were now to be distributed to this "select group of loyal, capable, and productive [*leistungsfähig*]" workers in order to promote the prospects of the most racially fit workers, to increase productivity in the factory, and to complement the mechanism of natural selection.[69]

Company Social Policy as Biopolitics

The new technocratic and racialized understandings of the relationship between work, reproduction, and the *Volkskörper* were linked to a number of institutional changes to the industrial workplace in Wilhelmine Germany. They were most clearly revealed in efforts to rationalize social provision in several of Germany's major electrotechnical, chemical, and iron and steel firms after 1900. This involved the creation of special offices and agencies and the hiring or appointment of new "social secretaries" for the administration of company benefits and social programs, which were often rearticulated in the languages of industrial and racial hygiene and materialized in practices and institutions that marked out three new domains of intervention: the biophysical capacities of the individual worker, the reproductive health of the worker's wife, and the physical well-being of the newborn working-class child. In this context, employer concerns began to (re)focus on the wives of industrial workers, their reproductive health, and their child-rearing practices.[70] By contrast with the primarily moralizing ambitions of factory paternalism, especially in relation to alleged moral benefits to be derived from "proper" familial organization, these new biomedical practices and institutions were designed to prevent the "squandering of human labor power" (*Menschenkräfte*), by (at least partially) reorganizing the conditions of industrial work and the procreative economy of the working-class family according to a wider biophysical calculus of labor productivity and racial "health."[71]

This emergent biomedical regime of company social provision emphasized the physical needs and functioning of the worker's body. In their programmatic social policy statement of 1906, officials at the Röchling steelworks announced that the most important task of company welfare was now the maintenance of a "healthy and work-capable [*arbeitsfähig*] core of employees."[72] Toward that end, they and other Saar employers began to institute a number of company programs and facilities related to personal hygiene and medical care. Indeed, the internal architecture of the Saar factory was largely reconfigured in the 1890s, as most industrial firms built a wide array of factory canteens, cafeterias, and special dispensaries for mineral water, coffee, tea, milk, and diluted beer; medical facilities and, in some cases, fully equipped hospitals; and a number of special convalescent homes and "invalid workshops" for "infirm and recuperating" workers. The Burbach steelworks and the Röchling concerns, for example, both replaced their small, relatively ill-equipped lazarettos with fully furnished, technologically advanced hospitals in 1896–97 and 1899, respectively. Indeed, the new hospital built by the Röchling concerns in Völklingen was soon expanded in 1908 and equipped with a surgical station, an x-ray room, a section for physical therapy, medicinal baths, massage rooms, and a sauna. These provisions were designed to meet the "bodily needs" of the healthy workers, the sick and injured workers, and the aged and infirm "in need of care" and who found themselves in a "condition of reduced labor power" (*verminderten Kräftezustande*) and no longer considered themselves of "value."[73] In addition, existing facilities were often upgraded and their functions reformulated in medical terms. In 1903, the Burbach steelworks, for example, despite the existence of two swimming pools, added a number of special bathing facilities—Turkish, carbonic acid, electric, and other medicinal baths, whose purpose was not only to cultivate good health but also to meet the "general needs of cleanliness and personal hygiene."[74]

The new biomedical calculus also suffused a wide range of other company educational or ideological initiatives. Officials at the Röchling steelworks in Völklingen devised a number of leisure and educational activities and institutions and recast existing ones for the purposes of sustaining the health and physical economy of the worker. In 1906, they created a "worker's library," equipped with "patriotic adult and youth reading materials as well as popular scientific books."[75] This was closely connected to the activities of the new company-sponsored yellow unions (discussed in greater detail in the next chapter), which entered into the realm of the worker's private life by means of pedagogical activities focused on the regulation of conduct in matters related to everyday behavior, economy, and medicine. Led by Hermann Röchling, one of the leaders of the national yellow union movement and Saar heavy industry after 1905, the Röchling yellow union sponsored guest speakers who delivered lectures on "economic efficiency" (*Wirtschaftlichkeit*), husbandry techniques for small animals, the dangers associated with the misuse of alcohol, and the everyday behavior that would facilitate healthy lifestyles or "hygiene" (*Gesundheitspflege*).[76] This was generally understood to promote the "capacity for performance" (*Leistungsfähigkeit*) of the individual worker. In a language that evoked the new technocratic calculus of efficiency, Dr. Karl Röchling made clear that the mission of the yellow unions was to disabuse workers of the notion that wages should be based on "legal or moral considerations" and to convince them that wages should reflect each worker's willingness and "capacity for performance."[77] The yellow union therefore became a critical site for the dissemination of the new biopolitical managerial rationality among their employees and larger efforts to disseminate ideas about national efficiency and the consolidation of the *Volkskörper* in the increasingly global economic struggle between nation-races.

Nevertheless, the most extensive new forms of Saar company welfare were designed to influence the procreative economy of the wage-earning family and the activities of working-class women in particular. In accordance with the biomedical practice of social-racial hygiene, Saar employers insisted that the "far-sighted employer who thinks about the future of his factory . . . must not only try to cultivate labor power [*Handkräfte*]; he must also support the women and the children [of the laborer] in order to cultivate the strength [*Kraft*] of the future generation."[78] They therefore developed programs that were designed to keep workers' wives at home, such as food transport services that relieved women of the task of bringing food to their husbands at the factory, as well as forms of social assistance after 1900 that simultaneously defined and targeted working-class women as mothers of future workers. The most important new programs involved medical care for pregnant women and assistance for women during the period immediately after giving birth. Company officials in Völklingen, for example, established an advisory office for expectant and new mothers (*Mütterberatungsstelle*) in 1907 and set up a specialized unit in the company hospital to provide immediate postnatal care for mothers. In addition, they created a program that offered in-home nursing and household assistance to women who had recently given birth. Managed through the local branch of the Patriotic Women's Association, caregivers were sent by the firm to the homes of their employee-clients to provide "care for the remaining children and the husband" and to maintain the household. The overall aim of this activity was to prevent "lasting damage to the health" of the mother from premature return to motherly duties, as well as to sustain the sexual division of labor in the household.[79]

Similarly, Saar firms began offering a variety of child care programs designed to improve the physical health of the "future generation." Thus, in 1904, the Burbach steelworks created a milk sterilization facility that provided nontubercular milk to families with newborn children. In addition, a maternity home, a crèche, and an orphanage were established after 1906 to provide care for infants and children of former employees. Officials at the Röchling concerns organized recreation activities for children of employees: their "summer break" (*Sommerfrische*) program involved an entire regime of outings, physical exercise, and nutritional provision for workers' children. From May to September, children were taken by coach out of the cramped and dirty industrial town of Völklingen and into the surrounding countryside, where they could increase their "body weight" and "regenerate themselves in the fresh forest air." "Under the influence of good forest air and [fresh] milk from summer to fall," according to factory physician Dr. Zillessen, "pale and suffering little mites are turned into rosy-cheeked, strong children."[80]

In addition to this emphasis on reproduction and child care, the new company social policy targeted the household economy of the working-class family and its sexual division of labor in a number of other ways. After 1900, several company welfare schemes involving education were devised—and others reconfigured—to promote what employers deemed to be efficient, orderly, and hygienic living arrangements among their employees' families. Most large steel companies in the region organized schools for small children (kindergartens) in the early years of industrial expansion. Around the turn of the century, however, firms began to create factory schools (*Werksschulen*) and apprenticeship workshops (*Lehrlingswerkstätten*) for the sons of industrial workers and sewing and home economics schools for the daughters. Teenage boys were offered instruction in a number of general skills, such as reading, grammar, simple math, geometry, drawing, garden cultivation, and gymnastics, and in various technical skills, including metalworking, carpentry, plumbing, bookbinding, brush making, and animal breeding. Generally speaking, the purpose of this kind of practical "economic instruction" was to provide for the "bodily development of young workers," "to ensure that the labor force will be appropriately trained, practically and theoretically, for its future occupation," and "to awaken in workers a sense for household efficiency, settledness, and thrifty living."[81]

The forms of schooling and "continuing education" offered to daughters of factory workers were also increasingly framed within the rhetoric of social hygiene. As early as the 1880s and 1890s, several Saar steel firms had created sewing and knitting schools for young girls, which were usually run by the local women's association (*Frauenverein*), headed by the wife of the factory owner or director. After 1900, however, the emphases in female education shifted to the development of home economics schools and the desire to promote a more hygienic and "efficient" working-class household economy.[82] To cultivate girls into "capable wives," the new housework schools at the Burbach steelworks, for example, offered courses in sewing, knitting, laundry and housecleaning skills, child care and discipline, cooking, and general housework. The 1906 test for a cooking class at the Burbach home economics school involved developing a fourteen-day plan to feed a family of four on a weekly budget of DM 11.54. The purpose was to demonstrate that a "tired husband and hungry children did not have to be fed only potatoes with coffee and bread" and that "a capable worker's wife could offer a variety of tasty meals with the aforementioned means." The teachers, of course, substituted several foods for meats, including puddings, vegetables, and some fish. Examinations at the end of each term were designed to cover "all of the duties of housework, cleanliness, orderliness, the preparation of a variety of meals, economical budgeting, and the correct use of income."[83] This emphasis on both "economic management" and "better nutritional maintenance" (*Ernährung*) was part of a larger ambition of the girls' schools to rationalize the gendered tasks of the household economy and the physical maintenance of the wage-earning family.[84]

Perhaps most revealing were the guiding assumptions behind the distribution at the Röchling steelworks of everyday "items of necessity," a provision formerly understood primarily in terms of moral charity and Christian obligation. According to company officials, if the principal aim of social policy was "to secure a health and work-capable core of laborers," it was necessary to develop a broad range of initiatives that would maintain the "good health of the population" and "above all . . . that of the future generation as well."[85] Toward this end, workers and their families were provided with everyday items such as food, heating coal, and clothing. Since many of the workers and their families were deemed incapable of attending to "matters of household economy," however, company officials and their assistants reviewed individual cases of need, dividing potential recipients into families of "inveterate drunks, idlers, etc.," on the one hand, and families who were hard pressed as a result of bad luck and other factors beyond their control, on the other. The value of these distinctions and "investigations of individual cases" was that they "revealed valuable clues about how the evils and defects [*Missstände*] could be remedied before they grew into dangerous diseases [*Volkskrankheiten*]."[86]

In this way, paternalist practice in heavy industry evolved into a new kind of "social medicine" (*soziales Heilmittel*). Designed to cultivate both a "greater willingness to serve" the company and healthy lifestyles among workers, this biomedical orientation involved managing from a distance, by means of the architectonic reconfiguration of factory space and the redesign of welfare programs from coercive paternalistic interventions to indirect biomedical and recreational services. It sought to regulate the physical health of the individual laborer and the entire laboring family as well as the working-class household economy in ways that would facilitate "healthy" social relations. In this way, it established direct links between the imperatives of factory production, the social and biological reproduction of the working-class family, and the racial health and fitness of the larger *Volkskörper*. This turn to biomedical considerations and the logic of bourgeois self-regulation partially displaced and reconfigured the directly coercive moralizing operation of company welfare schemes, along with their purported "personal" relations and intimacies, as the latter were previously deployed and understood within the framework of paternalist labor policy.

CHAPTER 6

Corporatist Discourse and Saar Heavy Industry

The new bioracial managerial rationality accompanied a wider transformation in the ideological discourse of employers and industrial culture. In a departure from the familial metaphors and representations long associated with the paternalist factory regime, many Saar industrialists began to reimagine work identities and social relations in distinctively corporatist terms during the decade before 1914. In their industry newspaper, journal publications, and internal reports after the turn of the century, they became increasingly preoccupied with a new "social aristocracy" of labor in the "productive economy" and a harmonious "community of work" in the large industrial concern. They also began to link these definitions of work and occupational identity to a larger social imaginary that articulated a corporatist vision of a world composed of "occupational estates" (*Berufsstände*). In this new ideological idiom, Saar employers began to call for the political organization of a wider "occupational estate of industry and trade" (*Gewerbe- und Handelsstand*) and the formation of a corporative sociopolitical order. The prewar ideological discourse of Saar employers therefore became corporatist in a dual sense: it articulated a worldview in a vocabulary that invoked forms of social address, natural hierarchy, and community that seem reminiscent of the corporate order of the old regime; and it formed the basis of programmatic political aims calling for the direct representation of economic interests in the realm of party politics and the state.

Scholars have long understood this corporatist discourse in terms of the persistence of premodern cultural and political traditions or antimodern sociopolitical developments in German political economy during the early twentieth century. Intellectual historians have traced corporatist ideas about social order to a specifically German tradition of romantic "antimodernism" that rejected "modern, liberal, secular, and industrial civilization" and helped pave the way for National Socialism.[1] Social historians have attributed the appearance of a residual and "romantic" corporatist ideology to the right-wing *Verbände,* which helped to forge the new economic and social interest groups of the Kaiserreich into a "secondary system of societal powers" that prevented the established political parties from coordinating competing social interests, frustrated the establishment of democracy, and strengthened the "traditional antiparty orientation of the authoritarian state" during the late Kaiserreich.[2] The practitioners of "social science history" formulated this interpretation explicitly in relation to the *Sonderweg* thesis, which associated corporatist representations of social order and calls for the creation of a "corporative state" (*Ständestaat*) with predominantly "preindustrial, precapitalist, and prebourgeois" social groups at odds with the political imperatives of modern industrial capitalism.[3]

Historical sociologists of corporatism have focused not on archaic ideologies and social relations in one nation but comparatively on the general structural realignments of advanced political economies during the twentieth century. In this interpretive framework, the term *corporatism* refers to a "new system of interest representation" common to the political economies of the twentieth century, involving state-sanctioned and regulated bargaining between noncompetitive and hierarchically constituted economic and social interest groups.[4] It permits analysis of the ways in which nearly all industrialized states in the twentieth century granted to various organized interest groups a "representational monopoly" over their specific fields of economic-social activity and displaced decision-making power away from the parliamentary arena toward the bureaucratized bargaining between state officials and large-scale interest organizations. Despite its illuminating comparative framework, however, this approach tends to portray right-wing corporatisms in places like "Fascist Italy, Petainist France, National Socialist Germany and Austria under Dollfuss" as state-centered projects derived from "delayed capitalist, authoritarian," and "neomercantalist" tendencies, rather as than "advanced" capitalist systems and their exponents;[5] and it rejects attempts to understand corporatism in terms of historically evolved political-ideological relationships as opposed to a universal "axis of development" in modern industrial societies.[6] By contrast, this chapter draws attention to the dynamics and productivity of discourse in the formation of social relations and the elaboration of capitalist political economies in order to offer a new interpretation of the generative context and forward-looking, productivist meanings of corporatism as a variant of capitalist ideology in the early twentieth-century Saar.

Corporatist Discourse in German Heavy Industry

Corporatist categories and assumptions shaped the theoretical reflection and social experience of a diverse range of philosophers, social theorists, state officials, and politicians in Germany and Europe during the nineteenth and early twentieth centuries. In contrast to the new corporatist theories about social order developed in France and Italy during the last third of the nineteenth century or just before the First World War,[7] German corporatist ideas appeared much earlier in the nineteenth century, in the political philosophy of Johann Gottlieb Fichte, Adam Müller, and Hegel. In the period after 1830, Catholic social theorists—including Franz von Baader, Bishop Ketteler, and Franz Hitze—embraced corporatist ideas about social and political organization as a response to the problems associated with industrial growth. In addition, estatist representations of social order were integral to the midcentury social theory of Wilhelm Heinrich Riehl, the traditionalist conservatism of Ernst von Gerlach in the 1860s and 1870s, the "monarchical socialism" of the prominent economist Adolph Wagner, the Christian Socialism of Adolph Stöcker, and the corporative experiments of Bismarck, who in 1880–81 attempted to establish a national economic council as a counterweight to the Reichstag.[8] Indeed, by the eve of 1914, corporatist terms and premises shaped the ideological visions of a range of political and economic interest groups, including the Conservative Party, the Agrarian League, and *Mittelstand* and peasant organizations. Capable of accommodating a wide range of signifying practices and programmatic aims, corporatism proved to be a multiaccentual discourse subject to different kinds of appropriation by a wide variety of historical actors.

During the Wilhelmine period, perhaps the most important (re)articulation of corporatist discourse came in response to intensifying labor conflict in German politics, as an effort to dissolve the unifying appeals of class. In the wake of the victories of the SPD in the Reichstag elections of

1903, the dramatic rise of trade unions and labor militancy, the formation of nationwide employers' antiunion organizations, and the growing "systematization of the weapons of struggle" at the workplace, in the form of coordinated measures such as regionwide strikes, lockouts, blacklists, and labor exchanges,[9] the editors of various industry publications and spokesmen for heavy industry began calling as early as 1906 for new organizations capable of protecting the economic interests of the productive or occupational "estates" (*Berufsstände*), as a response to the inroads made by the "class-political" trade unions and the SPD into what were deemed employer prerogatives at the workplace. After 1908, the growing popularity of corporatist vocabulary among industry publicists was unmistakable: key figures such as Hamburg factory owner J. A. Menck, Hamburg banker Max M. Warburg, Krefeld Chamber of Commerce chairman Otto Pieper, Düsseldorf Chamber of Commerce chairman Otto Brandt, and Saar industry spokesman Tille began publishing articles and pamphlets that explicitly referred to the imperative of occupational unity among German employers and elaborated plans for "occupational-political" (*berufsstands-politische*) organizations capable of representing employer interests.[10] Indeed, such initiatives—articulated in corporatist visions of a social order composed of "occupational estates"—increasingly came to define the social policy orientation and political strategies of the CVDI from 1908 to 1914.

The most systematic and tireless exponent of corporatist social theory within the ranks of German heavy industry was Tille.[11] In a series of essays and books—most notably his four-volume study of German political economy entitled *The Occupational Politics of the Industrial and Commercial Estate* (*Die Berufsstandspolitik des Gewerbe- und Handelsstandes*)—Tille argued that modern industrial society was composed of three principal groupings or "occupational estates" (*Berufsstände*): the estate of agricultural producers, the estate of educated professionals and officials (*Beamtenstand*), and the estate of industrial and commercial producers (*Gewerbe- und Handelsstand*). These social groups, organized by strict internal hierarchy, included everyone engaged in their respective areas of economic activity (i.e., agriculture, intellectual labor, and industry and commerce), from the wealthiest landowner, government official, and employer down to the lowest paid farmhand, clerk, or wage laborer. Indeed, Tille's "estate of industry and trade" comprised producers from twenty-one different branches of manufacturing and service industries, including all factory owners, managers, white-collar clerks, and manual laborers, along with all bankers, export manufacturers, merchants, retailers, independent artisans, and their employees.[12]

This was not, however, an atavistic representation of the social world, continuous with preindustrial schemes of social classification. Most versions of eighteenth- and early nineteenth-century corporatist vocabulary in the German states articulated a vision of a legally and religiously sanctioned social order of four hereditary estates, comprising the nobility, clergy, peasantry, and *Bürgertum.* According to this scheme, an individual's place in society was determined by "birth, official sanction, or ecclesiastical ordination" and was relatively fixed.[13] By contrast, Tille's proposed estates were consequences of recent economic growth, ordered along the lines of industry and occupation, and constantly evolving or "dynamic-meritocratic" (*dynameokratisch*) social forms. In his words, "the concept of estate had been transformed"; the former hereditary estates had been "superseded" by the current occupational estates, and the once "horizontal" arrangement of preindustrial estates (into upper, middle, and lower social layers) had given way to vertically aligned *Berufsstände,* each encompassing all levels of social status. Indeed, for Tille and other industrialists, the feudal connotations of the word *Stand* had largely given way by the Wilhelmine period to an entirely different understanding of *Berufsstand* as an occupation.[14]

This ideological rearticulation of corporatist terms explicitly rejected two other leading streams of corporatist discourse: *Mittelstand* protectionism and Catholic social theory. Tille rejected the corporatist social imaginary of the guild-bound artisan and all political demands based on the centrality of the small producer and the need to regulate and protect trade through state intervention. He criticized the way in which these corporatist claims and identities produced internal divisions within the wider "estate of industry and trade." In particular, Tille dismissed as "Marxist prejudices" the claims making their way "through broad sections of German *Handwerk*" that "economic activity that does not involve a certain amount of manual exertion constitutes idleness rather than economic performance and that its results, which appear in the form of profit and the increasing value of productive capital, were tantamount to unearned, immorally acquired, and illegitimate property." These assertions, he maintained, converged with the class-political "slogans" of socialist workers and failed to understand the central, generative force of "entrepreneurial daring and the entrepreneurial spirit" in contemporary economic life.[15] Similarly, the advocacy of Catholic social theorists in favor of a highly regulated and religiously grounded economic-social order of estates (*Stände-Ordnung*), as a substitute for the current order of an expansive industrial capitalism, was directly repudiated. Saar employers, under Tille's editorial direction, condemned such theories as grounded in an "ideal [based on] the compulsory guild of the Middle Ages," a violation of the freedom of trade as dangerous as Catholic attempts to limit the "freedom of conscience [*geistige Freiheit*]." The "order of estates" as described in such terms, they concluded, "is a fictional literary idiocy" but also "a sign of the fundamentally antagonistic stance that certain circles within *Ultramontanism* have taken against the liberal social order."[16]

As this reference suggests, the changed meanings of the term *estate* are particularly evident in the way Tille accommodated his corporatist sociology to certain principles of economic liberalism. An ardent supporter of what he called the "liberal societal order," in which the "freedom of the individual" was secured, Tille took pains to defend the "freedom of trade" and the "freedom of the wage contract," by which he meant the right of employers to set the terms of employment in their businesses without the involvement of the state, social reformers, or independent trade unions.[17] Tille staunchly defended the central features of the liberal economy, including private ownership of enterprise, private investment, and capital accumulation, which became subjects of a regular column entitled "The Liberal Societal Order" (*Liberale Gesellschaftsordnung*) in the pages of the weekly newspaper of Saar heavy industry during this period. This liberal framework, according to Tille, generated the conditions for a "dynamic-meritocratic society" (*dynameokratische Gesellschaft*): it released the creative capacities of the individual by offering "free room for maneuver to all forces" and "granted each the right to display his abilities . . . on the field of play."[18] In the realm of productive labor, Tille celebrated the role of competition between individuals and individual "achievement" or "performance" (*Leistung*) as the principal motor of industrial growth and progress. Precisely this "dynamic" economic framework and the industrial growth it fostered, he argued, had created the present social order of productive estates.[19]

In addition, as the previous chapter would suggest, Tille's corporatist social analysis, like his theories about the industrial workplace, derived from biological theories about human evolution and natural selection. Accordingly, his vision of a "dynamic" society of occupational estates rested on a social Darwinian paradigm that conceived of human relations in terms of an ongoing "struggle for existence," governed by the law of natural selection. Social relations within the occupational estates, he argued, were determined by the competition between biologically privileged or "capable" individuals, on the one hand, and those "less capable," on the other.[20] Work, as we have seen, was central to this design, because it was on the terrain of productive labor that individuals could realize their different innate capacities and abilities. The hierarchy between employers and workers in the world of production therefore reflected the "natural," biological inequality among individuals. A factory owner, according to Tille, simply possessed more evolved innate capacities, especially intellectual faculties, than a day laborer. Under these conditions, therefore, leading elements in the social order constituted a biologically determined (male) "social aristocracy" of ability and achievement in the realm of work and in the wider social grouping of the industrial and commercial "estate."[21]

In a way similar to his writings on the immediate domain of industrial work, Tille also defined the social aristocracy of work in the "productive economy" of occupational estates in relation to contemporary concerns and fears about the sexual division of labor and social and biological reproduction of the "body of the nation" (*Volkskörper*). In his theory of a social order of productive estates, as suggested in the previous chapter, Tille counterposed the masculine realm of competition in the productive economy with the male-headed (though largely female) household "savings economy," in which the "natural meritocracy of mental and physical attributes arranges itself . . . in a way that subordinates the woman to the will of the man and the children to the will of both parents." According to this scheme, the appropriate interaction between the savings economy and the productive economy would secure the biological reproduction of future laborers and promote austere household efficiencies that would produce biologically "fit" husband-fathers capable of competing in the realm of productive labor.[22] It provided the framework for a meritocratic productive and biological order that would secure favorable conditions for industrial growth and the health of the German *Volkskörper*. In this way, the maintenance of the health and the cultivation of the innate capabilities of biologically superior male individuals, who comprised the new "social aristocracy" of work in the social order of productive estates, would allegedly improve the economic and biological prospects of the German *Volkskörper* in the growing "struggle for existence" between nation-races.[23]

Company Unions and the "Community of Work"

The vision of a social order composed of vertical strata of occupational groupings also informed efforts to construct and symbolize new identities for workers, the most sustained of which came with the formation of yellow, or company, unions (*Werkvereine*) in the Saar beginning in 1906. Like employers in other German industrial regions after 1905, Saar factory owners, managers, and "loyal" employees created company unions in order to organize workers who would agree to reject strikes and membership in "strike organizations" (i.e., socialist, Christian, and liberal unions). The first such organization, the Burbach *Hüttenverein,* was created in late May 1906 at the Burbach steelworks in response to the organizational efforts of the CMV at the plant.[24] After the CMV launched its strike in June and began to extend its recruitment efforts to other industrial towns, officials at other large factories began to follow suit: company unions were subsequently created in 1907 at the Halberg steelworks in Brebach, the Röchling steelworks in Völklingen, the Röchling coking plant in Altenwald, the Vopelius glassworks in Sulzbach, and finally, in 1912, the Stumm steelworks in Neunkirchen. By means of coercive measures and incentives—most notably by linking all company benefits and social provision to membership in the union—Saar employers were very effective in organizing their workers in the new antistrike unions. In 1912, for example, the Burbach *Hüttenverein* numbered 4,671 members, 86.5 percent of the total labor force of 5,400 workers. In addition, they created an impressive regionwide movement by linking the unions together in the District League of Saar Company Unions in March 1912, which immediately became a corporate member of the national League of German Company Unions (Bund deutscher Werkvereine).[25]

Most historians view the company unions, modeled after similar organizations in the Ruhr and in Augsburg, as part of the long-standing paternalist strategy for controlling labor, but in the Saar, this turn to company unionism occurred in the midst of the growing crisis of paternalist hegemony and was linked to new efforts to redefine relations between employer and employee in the region after 1900.[26] This crisis was manifested in the rise of organized labor and the emergence of what company union organizer Dr. Karl Röchling called "class cleavages" at work all of which, he maintained, had rendered the paternalist vision of a familial and "personal relationship between employer and employee" much less convincing to a majority of Saar industrial workers.[27] Saar factory owners viewed the *Werkvereine* as a new organizational and ideological means by which to counter socialist images of the exploited wage earner and the factory as an arena of class struggle. The principal aim of the company unions was not simply to coerce the loyalty of workers or to seduce them with "material advantages" but to win their active consent to relations of inequality by offering them an alternative conception of the nature of work relations in the modern factory. Referred to by officials of the Röchling steelworks as the "inner conquest" of the labor force, this project was framed by a newly emergent corporatist ideological formation, which began to displace the paternalist model of work relations in favor of a new mode of interpellation in which the worker was identified as an independently minded employee, family father, and national comrade who rejected trade unions and sought to enter into a "peaceful" relationship with his employer.[28]

Under the leadership of Hermann Röchling, son of the founding director of the Röchling steelworks in Völklingen, and his social secretary Karl Rupp, Saar company union leaders set out to forge a new "solidaristic unity of entrepreneurship, capital, and labor"—a "community of work" (*Arbeitsgemeinschaft* or *Werksgemeinschaft*) in the factory or business enterprise.[29] They drew on the research and theoretical writings of Rostock economist Richard Ehrenberg, who, along with his student Erich Sperler, helped to create the prototypical company unions at the Siemens electrotechnical firm in Berlin and the Krupp steelworks in Essen and to train leaders of the *Werkverein* movement. The idea of a community of work was based not on the paternalist vision of a factory family or on preindustrial understandings of *Gemeinschaft* but, rather, on what its advocates deemed "scientific" and "empirical" observations of the nature of work and industrial organization in the age of machine technology and the large-scale factory. According to Ehrenberg, the increasingly complex division of labor and the vast differences in skill, income, and status in the "modern industrial enterprise" had given rise to a "community of work" in which employers, managers, and workers shared the same interest—the "business interest" (*Geschäftsinteresse*) of the firm—rather than to an arena of conflicting or competing interests. Yet the modern factory, unlike previous forms of craft production, had evolved to meet a vast array of local, regional, and global needs and therefore demanded far more complex organization (e.g., in terms of the division of labor and use of machine technology) and central direction. Consequently, trade union demands for codetermination and socialist demands for democratic control over the economy had to give way to the authority of the employer, whose managerial skills or "intellectual labor" became the decisive motor of factory production. Strict labor hierarchy, Ehrenberg maintained, became an existential imperative in the "community of work" of the modern factory.[30]

In contrast to the paternalist emphasis on the familial relations of the factory family and the employer as "provider," the rhetoric of Saar yellow union leaders emphasized a new combination of associational, hierarchical, and competitive aspects of the work community in the factory.[31] Local employers like Hermann Röchling began to address their workers, through the yellow unions, as "comrades" with shared interests, and the company union movement as a whole embraced quasi-egalitarian self-definitions and forms of address.[32] In his speech before a meeting of the Röchling yellow unions in August 1908, for example, the foreman Lober stressed both the equality and hierarchy among employer and workers.

> No matter what conditions we are born into, whether of high or modest rank, we are all men and we are all workers. The worker who gives us employment, however, especially deserves our highest respect, owing to his education and knowledge. To him often fall the difficult tasks that command very intense efforts and a great deal of knowledge. For this he earns our complete trust. We want to work with him—work with him to try to improve the situation of each individual.[33]

Company union leaders and functionaries therefore defined the legitimacy of their activities in distinctly populist terms, claiming that the new unions derived their authority from the support and desires of their wage-earning members rather than from the paternalist benevolence of the employer or the state. Röchling machinist Latz said, "We need no government saviors, like state secretaries or Obergenossen, to bring us good fortune." Indeed, company welfare schemes were vigorously redescribed: no longer deemed the product of employer charity (i.e., "alms" or "beggar's money"), they were regarded by the individual *Werkvereinler* as the legitimate "fruits of his labor," the just rewards for superior job performance. Indeed, the company unions sought to encourage this self-interested understanding of work and performance by means of combating the "leveling" tendencies of the trade unions—especially any attempt to establish a minimum wage catering to the "lazy"—and enforcing wage incentive structures that rewarded only the "diligent" and talented.[34] According to this self-image, company union members were deemed independently minded architects of their own fate, who had consciously decided to subordinate themselves to their employers. The company union leader Nikolaus Zimmer expressed this notion of "independence" with admirable succinctness: "We have our freedom just like the [other] trade unionists and we do what we think is right. If we cultivate a peaceful relationship to the firm in the yellow unions, then we are doing this out of the conviction that it is the best for us."[35]

Saar *Werkvereine* also developed organizational structures and activities that were designed to accommodate the everyday "living conditions" (*Lebensverhältnisse*) of their members and to cultivate the masculine and nationalist-racist ethos of the factory community—a new mode of interpellation that displaced the parent-child metaphors of paternalist discourse and recast workers as family fathers and national comrades. Most created administrative bodies and representative procedures characteristic of the voluntary association, including a governing council, a directorate comprising representatives from different factory sections, a general assembly of all members, and "courts" for settling disputes.[36] Beyond the formal meetings and administrative tasks, moreover, lay a much wider field of sociability that served to "cultivate a good comradery" and the "spirit of unity" among all employees and employers of a firm.[37] The yellow unions sponsored a broad range of educational and athletic activities, including libraries, lectures, musical and theatrical groups, sports clubs, and youth sections, as well as numerous social events (e.g., festivities and family outings), which were designed to provide for the leisure needs of workers and their families. According to Dr. Karl Röchling, these offerings and the general hostility toward trade unions best served the interests of the workers and their families.[38] In this context, the company unions defined workers as both independent employees and family fathers, who were now responsible for their own "dependents." In addition, the Saar company unions attempted to nourish the "feeling of brotherly community" among their members and to forge "comradely" bonds with other male industrial workers within the dense local network of nationalist associations. These ranged from choral societies and gymnastics clubs to the organizations more directly committed to propagandizing on behalf of German nationalism, *Weltpolitik,* and antisocialism.[39] Placing the Saar organizations squarely within the milieu of nationalist pressure groups on the German right during the prewar decade were their direct links with the local National Liberal party, the only viable "nationally minded" party in the region after 1900; their membership in the League of German Company Unions, of which Karl Rupp was the managing director; and their membership in the Council of National Workers' and Occupational Leagues of Germany (Hauptausschuss nationaler Arbeiter- und Berufsverbände Deutschlands), which included as members Hermann Röchling, Dr. Karl Röchling, and Rupp.[40] Finally, this national orientation was articulated to the new biomedical emphasis on physical health, social hygiene, efficiency, and "performance"—the incipient racialization of managerial rationalities—discussed in the previous chapter. Thus the Saar yellow unions, which offered lectures and reading materials to its members, became important sites for the elaboration and dissemination of the new biopolitical managerial rationality and for larger efforts to cultivate national efficiency and the consolidation of the "national body" (*Volkskörper*) in the increasingly global economic struggle between nation-races.

Finally, Saar employers, under Tille's direction, linked the community of work in the factory with efforts to build a larger "occupational-estate consciousness." Accordingly, the company union was deemed the best means of cultivating among workers the attitudes and values that would foster a "feeling of occupational community within the entire estate of industry and trade."[41] Employers viewed the yellow union as the ideal form of occupational association (*Berufsverein*), since it schooled its members in the laws of (corporatist) political economy and "economic peace." In the course of its meetings, educational lectures, and classes, the company union provided the necessary technical training for specific occupations in the factory, a "certain technical understanding for the enterprise and its operations," and a more general and scientific "economic knowledge" of the natural functioning and requirements of the industrial economy. Indeed, Tille himself was actively involved in organizing and delivering lectures on "economic and social questions" to the company unions.[42] Moreover, the Saar company unions incorporated ongoing efforts, discussed in the previous chapter, to identify the "skilled aristocracy of labor" in relation to an emergent "occupational consciousness" and more recent efforts to introduce new factory uniforms for long-standing employees at the Burbach and Röchling steelworks and the Fenne glassworks in 1905.[43] The new uniforms were modeled after the miners' uniforms, which consisted of the familiar black tunic, factory cap with plume, and leather apron, and were meant to invoke the corporatist traditions of the mining industry. At the Burbach steelworks, they accompanied the introduction of a new "factory festival," which also took the defunct "mining festival" as its model, and became the most visible symbolic representations of the elevated "corporatist" identity of the allegedly nonproletarianized company union man. "When the laborer wears his work smock with pride," Saar employers claimed, "he clings to his occupation and job with a different kind of passion than he would were he only a member of an undifferentiated multitude."[44] In this way, employers in the steel and glass industries hoped that the new uniforms would integrate the company unionist symbolically into the masculine, disciplined, and even martial formations of uniformed foundry men, whose physical strength, dexterity, discipline, and obedience to workplace superiors differentiated them from the (feminine) "'starving battalions,' with which Socialism parades."[45] In view of the strike threat and class conflict posed by socialist and Christian unions, these activities were deemed crucial to the ideological labor of imprinting "in the mind of every reasonably talented company union member a mental picture of the world of the economy as it really is today," easily grasped in a "conceptual system" and "slogans" (*Stichworte*) that would effectively counter the dangerous economic theory and "thieves' argot" of socialism.[46] "Much more modern" than the "class-political" trade unions, the company unions were designed to convince each individual member of the need to master the masculine skills necessary for superior job performance and for overcoming the "risks of life"—risks that were conditioned by "his health, his labor power, and his capabilities" and by the workings of a global industrial economy driven by large-scale factory production, productive occupations (*Berufsstände*), and the laws of racial-biological competition.[47]

The Organization of Labor Conflict, the Public Sphere, and the Corporatist Realignment of Industrial Interests

The previously discussed responses to labor organization and the increasing militancy of socialist (and, to a much lesser extent, Christian) trade unions were related to new efforts to establish nationwide antistrike coalitions among German employers and comprehensive attempts to organize all sectors of German capital into political pressure groups during the decade before the First World War. They were thus part of the more general corporatist realignment of German and European political economy after the 1890s. In Germany, this structural transition was propelled by the growing conflict between organized labor and capital after 1900. It emerged especially from efforts to forge an "employers' movement," initially as part of a response to the labor conflict in Crimmitschau in 1903–4, but also in relation to what industrialists perceived to be the privileging of organized labor and workers' interests in the main political parties, the Reichstag, the agencies of the imperial German state, and the "public sphere." These efforts at consolidation changed the scope and style of employer politics in Wilhelmine Germany, including the size and breadth of employer organizations and coalitions, the intended targets and audiences of their appeals, the kinds of functionary or "organic intellectual" directing their actions, and the ideological claims and discourses informing their strategies.

Locally, these changes were prompted by the determined efforts of Social Democrats and Christian trade unionists to break into the region after 1903, but the national impetus behind the formation of new industrial organizations and alliances among existing *Verbände* was the bitter textile strike in Crimmitschau from August 1903 to January 1904. In the midst of this strike, which was supported by the Social Democratic free trade unions, CVDI leaders collected funds in support of the Crimmitschau textile manufacturers. They then voted to create the Central Office of Employers' Associations in January 1904 and elected a committee led by Saar industrialist Richard Vopelius to draw up statutes for a new antistrike organization of German industry. On January 17, representatives from nearly all sectors of German industry met in Berlin to discuss the initiative, and negotiations took place between the leaders of domestic heavy industry in the CVDI and leaders of export-oriented and light and finishing industries of the League of German Industrialists (Bund der Industriellen, hereafter BdI). The CVDI-led initiative resulted in the formation in April 1904 of the Main Office of German Employers' Associations (Hauptstelle Deutscher Arbeitgeberverbände, hereafter HDA), which included representatives and organizations from the large-scale coal and steel, textile, paper, and glass industries. Despite efforts to forge a broad antiunion coalition, however, leaders of the BdI called for the formation of their own antistrike organization, subsequently named the Association of German Employers' Organizations (Verein Deutscher Arbeitgeberverbände, hereafter VDA), which was officially created in June 1904. The leaders of the VDA, which included industrialists and employer organizations from medium-size and smaller firms especially from the metal and construction trades, agreed with the antistrike ambitions of the HDA but were opposed to a leadership role of the CVDI and favored different kinds of antiunion strategies appropriate to the size of their firms and their different labor conditions.[48]

In the Saar, efforts to forge a broad antistrike organization for large and small industries were far more successful. In mid-May 1904, the leaders of all three industrial organizations voted to create a new regional antistrike association, the Employers' League of Saar Industry (Arbeitgeberverband der Saarindustrie, hereafter AVSI), which would collect funds in support of its members in the case of a strike and would support employers throughout Germany by means of its corporate membership in the HDA. Initially, its members were primarily from local private heavy industry, though until 1905, its initiatives were also supported by the Mining Office in Saarbrücken. In addition, Tille, as general secretary of the Saar organizations, actively sought to build a wider coalition with employers in smaller industries of the region, especially those firms—primarily in the brewing, construction, tailoring, bakery, and shoemaking trades—which were more susceptible to union demands and even had to tolerate socialist unionists among their employees. A spring boycott of the *Aktien Brewery* of St. Johann-Saarbrücken and a strike by local masons shortly thereafter prompted employers in the brewery and construction industries to seek new antistrike alliances across the region. In early March, the brewery owners of the Saar, Trier, and the Palatinate met with Louis Vopelius and Tille (who gave a talk about the national importance of these efforts) to create a new organization. After Tille wrote the statutes, the Employers' League of the Breweries of Southwest Prussia, the Palatinate, and Neighboring Regions (Arbeitgeberverband der Brauereien Südwestpreussens, der Pfalz und benachbarter Gebiete) was officially called into being on 27 April. Similarly, in mid-July, some sixty master artisans from the construction trades met in St. Johann to discuss the formation of an antistrike organization. In conjunction with Tille and the Saar Chamber of Commerce, they created the new Employers' League for the Construction and Associated Trades of the Saar Region (Arbeitgeberverband für das Baugwerbe und die verwandten Berufe der Saargegend, hereafter AVBS).[49]

Despite the split between the HDA and the VDA and the continuing tensions and differences between big domestic capital, on the one hand, and export-oriented and small capital, on the other, these efforts at coalition building signaled the beginning of a new era of employer politics during the prewar decade—a process crowned in any case by the "fusion" of the HDA and the VDA in the Alliance of German Employers' Leagues in April 1913.[50] This was evident in the organizational and ideological changes that transformed several of the leading industrial interest groups and their activities. First, German employers dramatically expanded the breadth and scale of their organizations during this period: the number of industrial interest organizations increased from 366 in 1900 to 3,670 in 1914; the latter included 3,040 local, 509 regional or district, and 121 national associations, though historians estimate that as many as four hundred or five hundred of all of these associations were active on the national level. Accordingly, the level of organization in a number of large-scale industries—namely, the extractive (49.7 percent), textile (50.5 percent), and metalworking (43.2 percent) industries—and the numbers of workers in their employ far outpaced the organizing capacities of the trade unions.[51] Second, this drive toward nationwide employer organization enabled new efforts to devise more effective antiunion measures, including more comprehensive blacklisting arrangements, lockouts, employers' labor exchanges, strike funds and insurance schemes, and the yellow unions. Third, industrial and other economic associations were transformed organizationally into more effective public pressure groups in ways that allowed new departures from previous lobbying practices. This involved the expansion of the central offices of new and existing *Verbände,* in terms of the size of their clerical staffs and the preeminence of a new "type of association politician" or "functionary" who took over the business of running the employers' organizations from the older-style practitioner.[52] Across most sectors of industry, this signaled the rise of a new kind of organic intellectual, whose primary responsibilities were to direct the organization and its ideological initiatives rather than to oversee or manage a particular factory or business. The Saar case is perhaps most illustrative in this regard: the former literature professor Tille, who built an entire public relations apparatus in the Saar after 1903, became the leading representative of heavy industry, replacing the factory owner Stumm, a parliamentary politician and self-described man of practice, who died in 1901.

The new centrality of the *Verband* functionary derived from employers' efforts to reorient their political activities more consistently toward a new audience and political domain, the "masses" as ideologically transformed into the public sphere (*Öffentlichkeit*) in the years after 1900. This important shift in strategies of interest articulation and representation both supplemented and departed from a long-standing practice of cultivating personal connections with leading ministers and government officials as well as delegates of the bourgeois parties. As in the Saar, the growing size and militancy of the trade unions across Germany and the electoral successes of the SPD—particularly the Crimmitschau conflict and the Ruhr miners' strikes of 1905—were in part enabled by and conducted on the terrain of *Öffentlichkeit.* German industrialists and their representatives recognized this fundamental transformation of public culture and sought to engage its discursive modalities and institutional forms. This explains the comments (cited in the introduction) of Eugen Leidig, deputy director of the CVDI, in the fall of 1905, about the two most important and thus new weapons of workers: the "universal and equal franchise" and "public opinion and the press," which, "with their silent and pressing authority . . . force all of our political parties to engage in the struggle for workers' votes."[53] It also indicates the seriousness of complaints by figures like Emil Kirdorf, Ruhr industrialist and member of the CVDI directorate, who acknowledged similar pressures and the deficits of heavy industry's existing channels of interest articulation, in his reflections on the role of the press just after the Ruhr miners' strike of 1905: "Unfortunately, I have come to the conclusion that we have not paid enough attention to the press and have underestimated its significant effects on the great masses (i.e., public opinion), parliament, and government."[54]

CVDI general secretary Bueck, who routinely invoked the factor of *Öffentlichkeit* in setting the terms and conditions of industrial relations and economic activity during CVDI and VDESI meetings after 1905, conceived of the HDA precisely in terms of its role in developing a "thoroughly active presence in the press" in order to influence public opinion. Despite the halting progress and mixed results of the CVDI's attempts to develop new strategies of publicity, its emphasis on "public relations" after 1900 was unmistakable and culminated in the massive press apparatus assembled by Krupp chairman and industry publicist Alfred Hugenberg after 1912.[55] But Gustav Stresemann and other leaders of the BdI, recognizing public opinion as a "power" but also as "driftwood" that was carried "here and there by various tendencies," sought most systematically to redirect its amorphous trajectories by means of "enlightenment"—"to win" public opinion over to the "justified interests of industry." "We must accept conditions the way they are," they maintained, "and say to ourselves: We live in the age of the influence of the masses, therefore industry must also rally the masses to it and attempt to exert influence through the masses on public opinion and legislation and on political parties." The latter adjusted their policies in relation to the "masses."[56] The central purpose of the BdI, according to its statutes, was to represent "the interests of German industry in relation to the claims of other branches of commerce and in relation to governments, parliaments, and the public sphere."[57] Like the Saar mining officials discussed in chapter 4, representatives of all branches of German industry increasingly acknowledged that the principal function of their organizational activity would be to influence public opinion in the public sphere in their battle against the trade unions, social democracy, social reform, and state *Sozialpolitik*.

This orientation toward *Öffentlichkeit* entailed efforts to establish new press organs as media of direct economic interest articulation in the formal political domain—a distinctively corporatist realignment of employer lobbying practices that meant abandoning the sole reliance on "political" or "neutral" newspapers under nominally independent ownership and editorial direction (including the *Kölnische Zeitung*, the *Rheinisch-Westfälische Zeitung*, and *Die Post*) to voice their demands. Leaders of the CVDI, which already published its own newsletter, were not very successful in this regard. By the end of 1900, they had secured control over the *Berliner Neueste Nachrichten*, which was to represent "the social policy and trading interests of German industry" within the framework of national interests and the "basic principles bequeathed by the founder of the Reich."[58] But criticism from the press, which focused on its direct links to heavy industry, led to the failure of this experiment by 1904 and a consequent reluctance directly to purchase another newspaper. As a result, CVDI leaders came to prefer strategies designed to secure their influence over existing newspapers via subventions and advertising. Nevertheless, other industrialists and their spokesmen, many of them members of the CVDI, were far more innovative. In addition to Stresemann, the most important early example of this reorientation came from J. A. Menck, who was president of the Altona Chamber of Commerce, a leading figure in the National League of German Metal Industrialists, and cochairman of the VDA. Menck was one of the most active of the new brand of industrial functionary and publicist; he was the leading force, along with editor Willy Reiswitz, in the establishment of the *Deutsche Arbeitgeber-Zeitung*, the first successful newspaper designed to represent the interests of German employers in the public sphere.[59]

In the Saar, where this turn to direct publicity occurred under Tille's direction, the corporatist political logic of these efforts was rendered explicit. While local industrialists renamed the industry newspaper *Neues Saarbrücker Gewebeblatt* as the weekly supplement to the *Neue Saarbrücker Zeitung* (hereafter *NSZ*) in 1896, only in the years after 1903 did the main press organ of Saar heavy industry evolve into an independent newspaper. This began with an editorial redesign under the name *Saarindustrie und Handel* in the summer of 1903. But since the latter remained under the control of the *NSZ*'s publishers, who were said to be "not able to treat the different interests that are represented in the papers that they publish equally" and to "consider the interests of private heavy industry last," local employers, along with Tille, established *Saarindustrie und Handel* as an "independent weekly" financed by the Saar Chamber of Commerce, the SGVDESI, and the VWGWISI toward the end of 1903.[60] This change, according to its editors, was part of a national trend involving the emergence of an "independent representation of interests" from within the ranks of German industry. In its final prewar format, introduced during the summer of 1906 and underwritten by all of the leading Saar business organizations, the *Südwestdeutsche Wirtschaftszeitung* (hereafter *SWDWZ*) was explicitly designed to redress the perceived failures of existing newspapers to represent the "pure" economic interests, unadulterated by moral and political considerations, of German industry and to describe their underlying "objective" origins.

> In the daily press . . . , whether it is representing the opinion of a political party or is completely neutral [*farblos*], the tone of economic interests in its complete purity only rarely appears. To be sure, there are also more objective and less objective papers, papers that print a certain range of perspectives, and papers that agree with a single party line, in the selection of their topics and in their general orientation [*Tendenzdarstellung*]. But an objective economic consideration of economic matters is offered only in the business sections of our major newspapers, which, however, almost daily produce the strangest contradictions with the opinions expressed in the political sections. Moreover, these business sections limit themselves almost exclusively to the reproduction of actual business news and overviews of the market situation. Insight into the causes of these conditions is only rarely attempted. Recently, the range of economic reporting in numerous daily newspapers has been considerably diminished by the fact that they have included moralizing considerations, alongside party-political views, instead of giving space to economic perspectives. For the

> most part, the former have a socialist tendency; after all, they seek mainly to agitate for the most equality possible [*Gleichmacherei*].

In order to combat the mainstream daily press, "sickened" as it was by a pervasive "social moralism," a belief in "equality," and a "one-sided concern with the interests of workers," the *SWDWZ* would represent the economic interests of southwest Germany and the putatively shared "interests and views" of all employers (*Unternehmertum*).[61]

This new orientation entailed an emphasis on the ideological production of corporatist economic interests and social identities in the activities of employers' associations. Not only were employers and their representatives increasingly concerned to (re)define the meanings of work, the worker, labor relations, and the functions of the modern business enterprise; they were now engaged in efforts to define the factory owner or employer (as "productive employer" [*gewerblicher Unternehmer*]), his interests, and his corporate relations to other social groups in the wider social order. This involved deploying the new definitions of the specific qualities and interests of the productive employer (understood as a unitary identity) in coalition-building initiatives within and between various sectors of German manufacturing and trade. Two of the most innovative centers of this kind of activity were Hamburg-Altona and the Saar, where industry representatives went about self-consciously forging wider unities among the disparate sectors of small and big capital. Menck and Reiswitz were the first industrial leaders and representatives to create a comprehensive employers' organization, the Employers' League of Hamburg-Altona, which included representatives from heavy to light industry as a matter of design and, in the pages of the *Deutsche Arbeitgeber-Zeitung,* attempted to rally employers to their own collective defense against the trade unions by means of the "cultivation of an esprit de corps" and the "consolidation" of their shared "consciousness of estate" (*Standesbewusstsein*).[62] It was precisely this model that they advocated and developed in the subsequent formation of the VDA.[63]

In a similar vein, Tille viewed the increasing coordination of employer efforts in the wake of the Crimmitschau strike—signaled locally by the creation of AVSI—as the beginning of an occupational-political "employers' movement," unifying a broad array of local, regional, and national coalitions among German employers into a formation of industrial-commercial concentration. Tille maintained that this was "a movement among much wider circles of productive employers . . . that goes well beyond circles of industrialists and therefore is better suited than just any industrialists' movement to form the basis of a political movement, even though it does not yet encompass trade and commerce."[64] Much like the Menck and Reiswitz initiatives in Hamburg-Altona, Tille's efforts to build contacts between a wide range of employers in the Saar and southwest Germany—through the chambers of commerce, the *Berufsgenossenschaft,* the brewers' and builders' associations, and the AVSI—were informed by the imperative to cultivate a shared "consciousness of estate" (*Standesbewusstsein*) among all "productive employers."[65]

Saar Employers, *Berufsstandspolitik,* and Corporative Sociopolitical Order

The ideological production of corporatist social identities and visions of corporatist political order were critical to the efforts to forge organizational links among employer organizations in response to what Saar employers perceived to be threats to their autonomy at the workplace and to their decision-making powers more generally—responses that brought about a striking ideological reorientation of employer politics in relation to state social policy, sympathetic political parties (especially the National Liberal Party), and the political-constitutional order of Germany itself. By contrast with their support for Bismarckian *Sozialpolitik* in the 1870s and 1880s, Saar and German industrialists associated with the CVDI now criticized the increasing volume of state welfare commitments, pursued by Secretary of Interior Posadowsky in particular: the expansion and restructuring of the invalidity programs in 1899, the expansion of sickness insurance in 1903, and the introduction of pensions for widows and orphans in 1911; new workplace safety measures and further restrictions on work time; the introduction of mandatory works councils in Prussian mining in 1905; and the government bill calling for worker chambers (*Arbeitskammer*) that was first mooted in 1905 and introduced in the Reichstag in 1908.[66] As chapter 4 suggested, the growing presence of trade unionists and Social Democrats in the governing bodies of the social insurance schemes proved particularly decisive in reshaping the attitudes of German industrialists. Accordingly, they began to locate the cause of these developments variously in the imperial government's reluctance to crack down on social democracy and the unions, the failure of the existing parties to defend employer prerogatives, and the universal male suffrage of the Reichstag, which allowed the SPD and Center to formulate labor and welfare policy. In short, employers began to criticize the entire party-political and constitutional system of imperial Germany, which they believed had subverted the vital interests of heavy industry.

In the Saar, employers decried these changes as threats not just to the racial order of the industrial economy but also to the "existential interests" of the "industrial and commercial estate." The accumulation of financial and regulatory burdens and the failure to curb the "extortion" of the trade unions, local employers maintained, had reduced "productive employers" (*gewerbliche Unternehmer*) to "citizens with lesser rights." According to this critique, German employers had increasingly become victims of trade union and "social-moralist" excess, of "ideologically construed claims for rights" that failed to comprehend the actual "forces" that sustained "economic and social life" in Germany. The "social-moralists" (i.e., social reformers, union leaders, and Social Democrats) ignored the central position of the employer as the superior "social aristocrat," the "bearer of the productive economy," who provided work, promoted technological innovation, and mastered the challenges of competition in the expanding industrial order—namely, harnessing the available "economic forces, productive forces, intellectual power, and labor power." The unbounded claims of "social moralists" therefore allegedly threatened the most important branch of German economic life—the "productive" workshops, factories, and businesses of the "industrial and commercial estate"—and the competitive prospects of German manufacturing in the international economy.[67]

These concerns fueled criticism not just of the dominance of the parties of reform (the SPD, the Center, and the left liberals) in the Reichstag, but also of the National Liberal Party, long an important bastion of industrial interests. In the years after 1900, as chapter 4 demonstrated, the left wing of the National Liberal Party, including its Young Liberal sections, began to push for a program of social reform and progressive taxation, forcing party chairman Ernst Bassermann to pursue strategies capable of keeping both left and right wings together. As early as December 1907, Axel Bueck, general secretary of the CVDI, had advised members of the VDESI to reexamine heavy industry's relationship to the National Liberal Party. Two months later, he warned of the leftward drift noticeable in the "social policy of the National Liberal Reichstag fraction that in many ways contradicted the interests of industry and employers."[68] At the same time, leading industry periodicals such as the *Deutsche Arbeitgeber-*

Zeitung were increasingly complaining about "insufficient representation" of industrial interests and about election to the Reichstag of party delegates "whose understanding of political economy amounted to nothing."[69]

In the Saar, the reorientation of the National Liberal Party, especially its embrace of the social policy and electoral strategy of its left wing and the latter's support for the striking steelworkers in Malstatt-Burbach in the summer of 1906, also generated growing dissatisfaction among local industrialists. Indeed, from 1908 to 1914, leading representatives of Saar heavy industry either abandoned the National Liberal Party or reexamined their commitments to it in light of continuing ideological affiliations and tactical electoral imperatives. Not surprisingly, the Free Conservative Tille, who was mistakenly ousted from the National Liberal Party for his remarks during the Burbach steel strike, organized much of this dissent.[70] In June 1908, he accused the National Liberals of abandoning their earlier support for the "ideals of political freedom" and the central tenets of Germany's "liberal social order"—that is, "freedom of trade," the "free wage contract," and the right of employers to set all the terms and conditions of work in business enterprises—and of supporting excessive social welfare and reform initiatives in their efforts to court "the masses during the Reichstag elections with promises of attaining heaven on earth for their voters." Speaking before a combined general meeting of the main Saar business organizations, Tille declared that the liberal parties were "so strongly infused with egalitarian ideas, indeed even communistic tendencies, that they have left the former ideals of political freedom in the background." The National Liberal Party's descent into the arena of social reform, he claimed, was simply part of a strategy to curry favor with the masses: "Out of National Liberalism has grown a National 'Equalism,' a National Socialism that will obviously transform one day into a National Communism."[71] In a formal statement, the executive committees of the SGVDESI, the VWGWISI, and the AVSI confirmed this perspective on the "privileging of the workers' estate" and their growing opposition to the National Liberal Party.

> A decade ago one could still call the majority of German industry National Liberal; today that is no longer the case. Since the National Liberal Party has embarked on a socialistic and egalitarian course in competition with the Center and social democracy over the support of the masses, industry has moved considerably further toward the political right. . . . The overwhelming majority of industry today refuses any connection to the National Liberal Party.[72]

This criticism continued in the employer press and found varying degrees of support among most of the leading National Liberal employers and their representatives in the Saar, including Weisdorff, Theodor Müller (director of the Stumm steelworks), Louis and Hermann Röchling, and Louis Vopelius.[73] Weisdorff, along with the National Liberal electoral association of his own district of Malstatt-Burbach, even broke temporarily with the local party organization in the spring of 1908, when the party supported the candidacy of the reformist Dr. Maurer for the upcoming Landtag election.[74]

In the years after 1908, these concerns about the National Liberal Party, coupled with the declining numbers of employers serving as delegates in the Reichstag, prompted a series of nationwide discussions about the political organization of employers. In the spring of that year, the Hamburg manufacturer Menck offered, in the pages of the *Deutsche Arbeitgeber-Zeitung,* the first widely discussed plan for a more effective employers' organization, a proposed "league of productive employers"—comprising all industrialists, artisans, and small shopkeepers—that could provide the lobbying strength employers needed to curb the social legislation of the Reichstag. Subsequent proposals for an employers' party, electoral organizations, and direct financing of campaigns were aired by numerous other representatives of heavy industry.[75] In addition, Tille called for the formation of a "league of employers" that would include producers in all branches of heavy and light industry, as well as large- and small-scale commercial, retail, and artisanal trades. Such an organization was needed, he maintained, in order to overcome the "divisive sectarian dispositions" among German producers—that is, to create the conditions in which "no one any longer thinks of himself as a member of a particular branch [of industry] but rather . . . as a comrade within an occupational estate." Rather than focus only on *Sozialpolitik* and limit itself to influencing the existing parties (like most of the other proposals), Tille's league was conceived most explicitly as a corporatist (*berufsstandspolitische*) organization that would replace the "obsolete" parties and represent employer interests directly in the organs of the state.[76]

Plans for a political organization of employers were finally realized with the creation of the Hansabund (Hansa League) in the summer of 1909. According to many German historians, the Hansabund marked the first comprehensive antiagrarian coalition of German capital before 1914 and thus embodied liberal aspirations for a politically active bourgeoisie ready to protect "commerce, trade and industry against attacks" from a "one-sided agrarian demagoguery" and to challenge the "feudal" power holders of the German state.[77] Yet many industrialists associated with the CVDI were also interested in the Hansabund. As several historians have pointed out, they regarded their participation in the new organization as necessary for tactical reasons—as a means of maintaining or resecuring the priorities of domestic heavy industry among economic pressure groups as well as their influence over this important new initiative in the "employers' movement."[78] But many leading German industrialists were also interested in the Hansabund as a corporatist form of interest representation. They conceived of it as the commercial-industrial version of the Agrarian League—that is, as an association that would represent the concerns of German industry during elections, in the parties, and at all levels of parliamentary government.[79]

In the Saar, this corporatist ambition was central to the self-understanding of the Hansabund's local branch. The initial July announcement for the founding meeting called on local businessmen to respond to the "attacks against trade, commerce, and industry" emanating from the Reichstag and to defend the "interests of the estate of industry and trade." It listed five general aims: the protection of "economic freedom" against onerous workplace legislation; the elimination of "one-sided" taxation policies at the Reich, provincial, and local levels; an end to "burdens" that were imposed on commerce, small business, and industry and favored other social groups; the prevention of "one-sided legislation" that privileged individual "classes"; and the maintenance of a balance between "duties and rights" in the social order and state.[80] Four days later, at the founding meeting in Saarbrücken, chaired by Richard Schmidt of the AVBS and attended by some eight hundred local businessmen, discussion of the new organization was largely shaped by corporatist categories and assumptions. Tille, who drew on his recent political writings to set the tone for the meeting, defined the Hansabund as the new representative of the interests of the "estate of industry and trade," which included employers in the building and artisanal trades, commerce and transportation, and mining and large-scale heavy industry. Its purpose, he maintained, was to arrest the declining numbers and political influence of businessmen-delegates or "productive employers" (*gewerbliche Unternehmer*)—in relation to the other *Berufsstände* (agricultural producers, white-collar workers, civil servants, and wage earners)—in the *Landtage* and especially in the Reichstag.[81] Tille's argument was followed by speeches about excessive state social policy, tax burdens, and the need for the self-organization of the "estate of industry and trade." The speakers included Arthur Olle, general secretary of the AVBS; master painter Wilhelm Schmelzer, speaking on behalf of the local *Handwerk;* and Weisdorff. The meeting ended with a unanimous vote on a resolution calling for corporatist forms of economic interest representation.

> The over eight hundred members of the Hansabund and supporters of its efforts who have assembled in the Saalbau in Saarbrücken on 23 July 1909 warmly welcome the political consolidation [*Zusammenschluss*] of trade, commerce, and industry in the context of the Hansabund. They are convinced that only a solid occupational [*berufsständische*] organization, which maintains its distance equally from party politics and confessional questions, is capable of establishing the political influence of the estate of industry and trade [*Gewerbe- und Handelsstand*] commensurate with its significance to the economic life of the people. Only by bringing the fundamental interests of productive employers to bear on all political parties and by sending more productive employers, proportionate to the number of their voters, into legislative bodies will the estate of industry and trade be able to guarantee in the long run legislation that will promote its fundamental interests and thus also the foundations of the economic well-being of the German Reich.[82]

In this context, corporatist discourse helped forge a broad unity of interests, under the direction of heavy industry, among a variety of businessmen and employers from both small and large businesses in the Saar—a coalition of employers that also drew on the linkages established by earlier efforts to build antiunion organizations and funds throughout the region. The initial announcement was signed by local employers from the ranks of the artisanal trades, the construction industry, small retailing, banking, and heavy industry, including most of the politically active industrialists and their representatives (Tille, Weisdorff, Louis Röchling, Theodor Müller, and Louis Vopelius) and all of the leading business organizations (the VWGWISI, the SGVDESI, the AVSI, and the Saar Chamber of Commerce).[83] The first meeting of the Saarbrücken chapter produced an executive committee comprising Schmidt from the artisanal trades, businessman Heinrich Simon of Sulzbach from commerce, and Weisdorff from heavy industry. A similar occupational parity was achieved in the election of the 150-member governing council, which drew forty members each from *Handwerk,* commerce, and heavy industry; its remaining thirty members were chosen from among the ranks of local white-collar workers—that is, from the managerial and clerical personnel employed by local firms. By early November 1909, the Saarbrücken chapter of the Hansabund reported a total of eleven hundred members. Nevertheless, the dominance of heavy industry was secured in a number of areas, ranging from the finances available to the new organization, which mostly came from heavy industry, to the influence of Tille, who, as the managing director of the local chapter, was responsible for running its day-to-day affairs and representing its interests in the governing council of the national organization.[84]

Yet the articulation of employer economic interests within this corporatist understanding of the Hansabund had little to do with what historians have long referred to as traditional, anticapitalist, or "feudal-conservative" conceptions of social order. Nearly all of the speakers at the first meeting of the Saar branch criticized government social and taxation policies not for their failure to sustain or promote "preindustrial" business practices or feudal models of economic organization but, rather, for their violation of the "free wage contract" and their subversion of the putatively natural relationship between wages and job performance. Far from demanding protectionist legislation and legal privileges, they claimed to be supporting "economic freedom" or "free trade," employer "rights," and entrepreneurial initiative. In his speech on behalf of small shopkeepers, for example, Wilhelm Weiten defined the "tradesman" in terms that resonated with the masculine, capitalist self-image (long propagated by Tille himself) of most businessmen and industrialists: "The business owner [*Gewerbetreibender*] . . . is an employer who relies on his own initiative; an independent man, who has his own business, who is paid [a wage or salary] by no man and cannot be dismissed from employment by anyone, who is integrally connected to his business, from which he incurs huge risks but also, if he is capable, from which he earns a profit." Indeed, the rhetoric of local Hansabund leaders bears no trace of precapitalist, antimodernist ideologies of "economic despair" long identified by historians; rather, their claims and aspirations were expressed in an ideological framework that was self-consciously "modern" and forward-looking and grounded in a vision of economic progress.[85] Thus Tille reiterated his claims about the generative activities of "productive employers" and the "estate of industry and trade" as the central *Berufsstand* of the new industrial era, which "under its own steam would steer its own course toward its own goals"; and Olle addressed the role of the Hansabund in bringing about a new era of prosperity for German business as the result of "the old, glorious German Hansa of past centuries" being "reborn in modern guise."[86]

Perhaps most telling, this new corporatism distanced the political demands and self-definitions of the *Gewerbe- und Handelsstand* from the policies and ideological orientations of the protectionist agrarians and *Mittelstand.* In this regard, Weisdorff clearly articulated the opposition of heavy industry to all "enemies of economic life," either from the "left" or the "right." If the former brought the "suffocating pressure toward equality," the latter, he argued, threatened economic expansion in an era of global markets with a stultifying "traditionalism"—the economic self-limitations intrinsic to the "eternally conservative." In between these two positions, Weisdorff continued, "lives freedom of trade, economic freedom, the freedom to realize . . . the value of one's mental and bodily abilities in useful creation, in free labor."[87] From the perspective of *Mittelstand* aims, Weiten explicitly contested the economic demands and the very schemes of social classification of the "so-called *Mittelstand* movement." In a sustained reflection on the "occupational estate" of productive employers, he rejected attempts to locate small retailers, artisans, and *Beamten* within a single *Berufsstand,* but he also rejected the category *gewerblicher Mittelstand* for its arbitrary closures.

> The productive *Mittelstand* is also nothing special in and of itself. Who wants to say where it begins and where it ends? Why should the smallest and the largest employers be excluded from it? There are no fixed boundaries. Whoever is a small business owner today will perhaps already be a medium-sized business owner in one year and in ten years a large business owner; and it is not true that there are burdens that hamper only small business or only big business. In the long run, all have to bear these burdens.[88]

Not surprisingly, Tille's industrially oriented, dynamically capitalist, and self-consciously modern variant of corporatism dominated discussions of the local branch and its coalition-building efforts, publications, and educational courses, for which Tille's four-volume *Die Berufsstandspolitik des Gewerbe- und Handelsstandes* served as the textbook.[89]

Indeed, the failure of the Hansabund to conform to this corporatist vision eventually generated local opposition to its national directorate. Saar employers accused the national organization of failing to represent the occupational interests of the "industrial and commercial estate." This was evident in its neglect of the latter's economic concerns—particularly strategies to reduce the amount of social legislation and to enact legal measures to curb trade union activity—and also in its pandering to political parties of the Left.[90] Moreover, the Hansabund was conceived in the Saar as a "occupational-political" organization of the "estate of industry and commerce" rather than a political party or an economic association with political aims. As such, it was supposed to represent the economic interests of the entire estate of industry and trade in the parliamentary realm. According to Tille, the Hansabund failed to do this in two ways: it neglected the economic interests of the productive *Bürgertum,* and it failed to embrace all members of the occupational estate—namely, the white collar and industrial workers who rejected the class-political aims of the labor movement and the "native" (*bodenständig*) and patriotic workers of the yellow unions.[91] For these reasons, Saar employers, along with other representatives of heavy industry, withdrew their support for the Hansabund in the summer of 1911.

In the final years before 1914, these frustrated ambitions—compounded by the results of the Reichstag elections of 1912, from which the SPD emerged as the largest party in Germany—prompted German industrialists to pursue their corporatist commitments in other ways. Within the "occupational estate" of industrial producers itself, the leaders of German heavy industry began to help forge nationwide organizational links between the CVDI and the yellow union movement. In 1910, the latter created the aforementioned League of German Company Unions and shortly thereafter joined with the League of Patriotic Workers' Associations—a national confederation of workers' organizations that were not specific to any company or industry—in the Council of National Workers' and Occupational Associations. The council eventually organized an antisocialist alliance of laborers from a variety of trades and industries, including factory workers, artisans, retail workers, and sailors.[92] To strengthen this "bulwark against the red danger," the CVDI established official ties to the council in December 1912, lending publicistic and financial support to the company and patriotic unions in the final prewar years. In the Saar, similar ties were established between the Employers' League of Saar Industry and the regional federation of company unions. These were aimed at forging "a closer union among all employers, managers and clerical workers, and wage laborers of the productive economy," to help "foster the feeling of occupational consciousness within the entire industrial and commercial estate."[93]

At the same time, the CVDI began reinvigorating its efforts to win the support of the *Mittelstand* of small producers. In the summer of 1911, CVDI leaders joined with representatives from the Agrarian League to assist in the creation of the nationalist Imperial-German Mittelstand League. Promoting the alliance of "*Handwerk* and industry," according to CVDI chairman Max Roetger, was necessary to halt the steady advance of social democracy and social legislation that subverted employer interests. Tille, who became the leading advocate of *Mittelstandspolitik* within heavy industry, cast this initiative in distinctly corporatist terms. In his programmatic article entitled "The Industrial League" and in his speeches before the Imperial-German Mittelstand League in 1912, he stressed the importance of small business to the unity of the industrial estate. Small workshops and commercial businesses, he argued, were equally burdened by the social legislation that privileged wage earners. Much more serious, however, was the growing power of Marxist "prejudices" widespread within the small business community itself. These included "moralistic" criticisms of entrepreneurs as "robber barons" and hostility toward capitalist forms of investment, profit making, and managerial-entrepreneurial (or nonphysical) labor. A broad-ranging coalition or "industrial league" involving producers from the mining, manufacturing, commercial, transportation, housing, retailing, and artisanal sectors therefore would help root out socialist ideas in the *Mittelstand,* consolidate the occupational unity of the entire estate of industry and trade, and defend the capitalist industrial order.[94]

These corporatist initiatives were central to the wider process of ideological realignment taking place on the German right. During the final prewar years, representatives from German heavy industry and agriculture attempted for the first time to forge a genuinely populist coalition, by actively appealing to the interests of small-scale producers of the *Mittelstand* and the peasant classes on the basis of a radical-nationalist appeal.[95] By the spring of 1913, a wide range of groups were actively pursuing organization contacts, including the CVDI, the Agrarian League, the Christian Union of Peasants' Associations, the Imperial-German Mittelstand League, and several nationalist pressure groups. The main representatives from heavy industry pushing for this realignment were the Ruhr industrialists grouped around Alfred Hugenberg and Max Roetger of the CVDI, Emil Kirdorf of the Rhine-Westphalian Coal Syndicate, and the Saar industrialists led by Alexander Tille (until his death in 1912) and Louis Röchling.[96] Their goal was to forge a common political alliance organized around commitments to antiunion legislation (especially laws against picketing), a cessation of state *Sozialpolitik,* antisocialism, and German imperialism. Finally, representatives of all of these interest groups, along with leaders of the Pan-German League and the Imperial League against Social Democracy, met in Leipzig to announce the formation of the Cartel of Productive Estates in August 1913. As its program suggested, the principal tasks of this alliance of the main productive "estates" or occupations were to unify industrialists, agrarians, and *Mittelständler* in defense of their economic interests; to maintain "authority in all economic enterprises"; to secure price guarantees and the "protection of those willing to work" (strikebreakers); and to defend against social democracy and "socialist heresy."[97] In recognition of the "common existential interests of industry, *Handwerk,* and agriculture," according to CVDI secretary Ferdinand Schweighoffer, the cartel emphasized the "necessity of these three *Berufsstände* standing side by side in all fundamental questions related to our economic and social policy."[98]

In addition, the cartel's call for the formation of an "economic parliament" with decision-making powers in the areas of economic and social policy revealed the growing interest of key sectors of heavy industry in a state-sponsored coup d'etat or in new corporatist strategies for bypassing the existing parliamentary and constitutional order altogether. The strongest public signal for this commitment within domestic heavy industry came from Max Schlenker, who replaced Tille as head of the Saar Chamber of Commerce after the latter's sudden death in 1912. In his essay of June 1913 entitled "Revision of the Reichstag Suffrage or the Creation of an Imperial Upper House?" Schlenker criticized recent debates in the Reichstag over the Army Bill, which proposed direct taxation in the form of business and property taxes. He warned that the Reichstag had "embarked on a path that must fill the [productive] estates with the utmost concern about the future." In response to this most recent invasion of the Reichstag into the realm of private property and to the deeper problem of the dearth of businessmen-delegates in the Reichstag, Schlenker proposed two possible constitutional solutions: (1) changes in the electoral laws (e.g., property qualifications and residency requirements) to guarantee the number of businessmen delegates in the Reichstag and (2) the creation of an imperial upper house composed of representatives of the occupational estates (*Berufsstände*). According to Schlenker, only constitutional change could prevent legislation from restricting further the "discretionary power of the employer in his factory" and could shield this "aristocracy of labor" from the tyranny of the "unpropertied majority."[99]

Schlenker's call for an imperial upper house and electoral reforms found growing support among German industrialists. Indeed, growing hostility toward mass politics spawned a number of possible schemes for constitutional revision in 1913, emanating from the Ruhr and Saar in particular and published in the leading press organs of heavy industry.[100] Yet demands for property qualifications to existing imperial and Prussian suffrage laws and designs for the creation of an upper chamber that would act as a countervailing force to the Reichstag found a growing audience among even wider sectors of German heavy industry. They were raised in the context of discussions within the CVDI and in the *Deutsche Handelstag* (the collective forum of the chambers of commerce), and by 1913, these goals, in more limited forms, were winning adherents from the ranks of the export-oriented and liberal free-trading firms of the Hansabund, the Association of Saxon Industrialists, and the BdI.[101] The general sense of domestic crisis within the ranks of German industrialists in the wake of the Social Democratic electoral victories of 1912 and fears about competition in the global economy therefore rendered corporatist ideological appeals increasingly attractive to growing numbers of representatives from the finished goods industries and smaller-scale manufacturing. In this way, corporatist discourse contributed to the ideological convergence between employer organizations and the nationalist pressure groups of the radical Right after 1913.

EPILOGUE

Toward a Genealogy of Fascist Corporatism

The emergence of corporatist discourse in the Saar was in some ways distinctive to the region, especially in terms of its relatively systematic theoretical elaboration, but it anticipated the general corporatist turn in the politics of German and European industrialists and radical nationalists in subsequent decades. In Germany, it was redeployed in relation to the growing interpenetration of public and private power during the First World War, during which new structures of interest representation and economic planning that involved employers, workers, and state officials emerged; in opposition to the formation in November 1918 of a "central works community" (*Zentralarbeitsgemeinschaft*), which brought employer associations and trade unions together in a common system of collective bargaining and state arbitration; in opposition to the role of state officials and parliamentary politicians in the formulation of national economic policy in relation to tariffs, wages, finance, currency, and foreign trade during the Weimar Republic; and in response to the crisis of productivity and unemployment brought on by the global economic collapse of 1929–30. Indebted to fixed ontological distinctions between ideology or discourse and the economy, as well as between the state, the public sphere, and the industrial workplace, historians have largely overlooked the productivity of this version of right-wing corporatism: namely, the capacity to assemble a disparate array of linguistic resources into a new ideological configuration, which articulated racist visions of a social order organized by self-administering "occupational estates" to a modern rationality of industrial management; and its capacity to effect changes to the social relations of the industrial workplace and the wider domains of industrial interest representation and employer politics. We must look to these capacities if we are to appreciate the ideological appeal of corporatist discourse within the ranks of German heavy industry after the First World War. During the 1920s and 1930s, corporatism proved to be an important field of ideological debate and convergence between the leaders of German heavy industry and the National Socialists—a convergence that brought them together in their attacks on the Weimar Republic during the early 1930s and in the subsequent elaboration of the Nazi project after 1933.

Rearticulated versions of the new managerial rationality and "scientific management" explored in this study were taken up in new institutes of industrial sociology and social policy, especially the Ruhr-based German Institute for Technical Labor Training (Deutsches Institut für technische Arbeitschulung, hereafter Dinta) in Düsseldorf. Established in 1925 by Ruhr industrialists (especially Albert Vögler of the Vereinigte Stahlwerke) and led by the engineer Karl Arnhold, Dinta developed new labor policies and training programs and disseminated information about schemes of industrial rationalization, "human economy" (*Menschenökonomie*), "human management" (*Menschenführung*), and company social policy, with the aim of reaching employers throughout German industry. It sought to transform the "soul of the worker"—to cultivate obedient workers, imbued with the virtues of military discipline and a belief in social hierarchy, who would reject class-based trade unions, socialism, and communism and come to celebrate their "joy in work." This ideological project centered on efforts to induct the male worker into the world of individual "performance" (*Leistung*) on the job, loyalty to the collective "works community" (*Werksgemeinschaft*), and pride in one's occupational "estate."[1] It also entailed the biopolitical disciplining of the industrial workforce, which meant the introduction of gendered workplace schemes and measures related to social and racial hygiene and extending to workers' familial lives off the job, including exercise and sporting activities, nutrition and health care provisions, schools for household pedagogy and rationalization, and services for maternity and infant care.[2] By 1930, Dinta was active in nearly three hundred German and Austrian firms, mostly in the iron and steel, metalworking, and mining industries; its ideological terms—references to the workers' "estate," the "works community," and job "performance"—suffused the political language of German industrialists, who sought to destroy the system of collective bargaining and labor arbitration during the Weimar era.[3]

These definitions of the *Werksgemeinschaft* were generated in postwar discussions that brought together industrialists and their representatives and various politicians and spokesmen from the right-wing organizations and political parties, including the Nazis, who invoked a corporatist conception of workplace organization as part of their larger plans for the "creation of a corporate social order" (*ständische Aufbau*).[4] But it was mostly the Dinta model, emanating from the ranks of heavy industry, that was integrated into the labor and social policies of the Third Reich. The institute, including its director Arnhold, was seamlessly "coordinated" into the Nazi system after it was renamed the Office for Factory Leadership and Occupational Training and was taken up in the activities and programs of the German Labor Front (Deutsche Arbeitsfront, hereafter DAF), the main social policy organization of the Nazi era.[5] Most of Dinta's principal ideological terms and demands, especially its emphasis on "performance" and the "works community" (*Betriebsgemeinschaft*), were written into the labor law of the Third Reich and institutionalized in Nazi wage policies. Indeed, during this transition, representatives of German heavy industry successfully lobbied on behalf of a corporatist vision of "occupational estates" that would secure the prerogatives of the employer in the workplace and institutionalize their scheme for relative employer autonomy and industrial "self-administration" in relation to factory social policy, and they were the principal authors of the Law on the Organization of National Labor of 20 January 1934.[6] For their part, the Nazi leaders of the DAF reformulated their corporatist plans for *ständische Aufbau* in relation to the industrial workplace during the summer and fall of 1933, in ways that accommodated demands for autonomy and employer prerogative emanating from the ranks of German heavy industry.[7]

After 1933, corporatist references to occupational "estates" and the "works community" combined with racial concerns about the worker's body, individual "performance," and the social and biological reproductive functions of the working-class family to shape ideological discourses about work and newly reconfigured factory regimes during the Nazi era. Bioracial corporatist terms and assumptions figured centrally in the writings and policies of Nazi physicians and scientists, industrial sociologists and efficiency experts, and DAF functionaries; and they constituted the core of racist definitions of work, occupational hierarchies, and social order in the much-vaunted "performance community" (*Leistungsgemeinschaft*) of the Third Reich.[8] This was visible, for example, in the rationalized industrial workplaces of the automotive firm Daimler-Benz, at the electrical firm Siemens, and in the plans for the National Socialist model factory of the Volkswagen concern, which was to introduce rationalization measures for its main plant in conjunction with the formation of a new "worker elite" created by means of vocational training and a company policy of bioracial "selection."[9] These racial distinctions and social policies became especially important to the "factory community" during the Second World War, when German industrialists and managers incorporated millions of "racially" foreign slave laborers into the Nazi economy.[10]

The prewar "corporative antiparliamentarism" of German industrialists proved equally consequential for the history of labor relations and parliamentary politics during the Weimar era. As Dirk Stegmann demonstrated, the corporatist demands and experiments proposed by Saar and other German industrialists, especially in the context of the Cartel of Productive Estates of 1913, were revived during the First World War. Their leading exponents were among the organizers and supporters of various initiatives related to the formation of a German-dominated customs union in Europe (*Mitteleuropa*), often orchestrated by leaders of the Pan-German League, and to the creation in the fall of 1917 of the German Fatherland Party, which attempted to remobilize a broad-ranging coalition among the forces of the political Right on behalf of annexationist and imperialist war aims and to counter threats of democratization in Germany.[11] The industrialists and industry representatives who supported the party, especially leaders of the CVDI, were particularly concerned to resist demands for an end to the three-class voting system in Prussia emanating from the Reichstag and to bring about constitutional changes in the direction of a corporatist political order, including an "economic parliament" and a "corporatist electoral law" organized by productive estate.[12] The center of this activity shifted to Rhineland-Westphalia after the war and revolution, when plans for corporatist economic and social (re)organization were most systematically debated among industrialists in the Ruhr, even though its influence was curbed during the mid-1920s by the accommodationist line adopted by the new umbrella organization for German employers, the Reichsverband der Deutschen Industrie (hereafter RDI), and its leader Paul Silverberg. As early as May 1920, Vögler, head of the VDESI, was disparaging the newly created parliamentary order in favor of schemes of "economic self-administration" for the resolution of "all economic questions by means of a corporatist [*berufsständisch*] economic parliament."[13] Subsequently, these discussions took place most frequently in the meetings of the Langnamverein, the main employers' organization for the Rhineland and Westphalia, and in the meetings and congresses of the League for a National Economy and Works Community (Bund für Nationalwirtschaft und Werksgemeinschaft), a radical right-wing pressure group created in 1926 under the leadership of the Pan-German Paul Bang, a close ally of Alfred Hugenberg and Vögler. Leaders of this latter organization combined racist definitions of economic competition and industrial management, demands for the "self-administration of the economy" by representatives of the leading productive estates, and attacks on the "Marxist social policy" and parliamentary government of the Weimar Republic in a "modern" capitalist vision that emphasized the virtues of a "free economy."[14]

This articulation of an "estatist" social imaginary to a technocratic capitalist rationality characterized many of the right-wing corporatist projects across Europe during the 1920s and 1930s. Corporatist theories developed in all European countries during and after the First World War, when powerful pressures for economic regulation and planning, which involved new state institutions and ministries alongside organized industrial interest groups, were created as the means by which to order the chaotic dynamics of industrial capitalism.[15] During the 1920s and 1930s, radical nationalist and conservative groups put forward their own corporatist schemes for industrial social order, which were variously implemented in Italy, Spain, Portugal, France, Austria, and Germany. Far from being the backward-looking, restorationist, or even anticapitalist ideological projects that appear in many histories of corporatism, the most important right-wing versions of corporatism, especially their authoritarian and fascist variants, were complex ideological formations, which generally proposed forward-looking schemes for the creation of producers "estates" or chambers and varying measures of state planning that included visions of employer prerogative and "self-administration" of trades and industries, the reinforcement of capitalist property relations and markets, and autarkic—though not isolated—regions of economic development. These corporatist schemes were forged in wide-ranging political debates between fascist politicians and intellectuals, fascist employers, and conservative industrialists and their representatives—debates that involved demarcating the corporatist modernisms of European industrialists from the dirigiste, protectionist, and "guild" models of corporatism favored by small-scale producers and retailers or the industrial bargaining models of corporatism favored or encouraged by certain labor leaders and reformist employers. This was the case with French industrialists, who favored legal recognition of their own cartels in the 1930s; Spanish industrialists during the Francoist reorganization of the economy during the late 1930s; and Italian industrialists and nationalist corporatists like Antonio Benni and Alfredo Rocco, who successfully outmaneuvered supporters of a worker-centered syndicalist corporatism in order to establish a corporatist framework of industrial production that favored the authority and interests of heavy industry during the late 1920s and early 1930s.[16]

In Germany, corporatist modernism competed with other variants of corporatism during the 1920s. By the early 1930s, it became the principal ideological discourse through which representatives from domestic heavy industry criticized the system of state labor arbitration, social policy, and constitutional structure of the Weimar Republic. Most historians of corporatist ideology and Nazism have focused on the corporatisms of small producers (German retailers and shop owners), with their calls for guild-like organizations and guarantees of state protection. Indeed, Hitler and several other Nazi leaders—including Gottfried Feder, Otto Wilhelm Wagner, Gregor Strasser, and Max Frauendorfer—were attracted to dirigiste corporatist solutions to class conflict and the volatility of capitalist economies, especially to the "neoromantic" theories of the Viennese professor of political economy Ottmar Spahn.[17] Nevertheless, Spann's theories left much more scope for the self-organization and autonomy of German industry, and the Nazis proved willing to drop some of their seemingly anticapitalist positions as early as 1926 and 1928.[18] Moreover, small producers' demands for corporative organization of the economy competed with corporatist theories that were firmly anchored in capitalist rationalities, most notably Walther Rathenau's and Wichard von Moellendorff's blueprints, developed during the First World War, for a "German collective economy" (*Deutsche Gemeinwirtschaft*), as well as the corporatist strategies emanating from the ranks of heavy industry, especially in the Ruhr. From 1930 on, German industrialists and their representatives debated the latter in the meetings and congresses of the leading industrial organizations, including the Langnamverein, Bergbau-Verein, RDI, and Deutscher Industrie- und Handelstag, as well as in their own newspapers and newsletters, including *Der Arbeitgeber,* the *Mitteilungen der Vereinigung der Deutschen Arbeitgeberverbände,* and the *Deutsche Führerbriefe,* a newsletter connected to heavy industry via the Langnamverein and the Central European Economic Congress (Mitteleuropäischer Wirtschaftstag). The principal exponents of corporatist industrial social order within the ranks of heavy industry included Albert Vögler, Max Schlenker, Fritz Thyssen, August Heinrichsbauer, Karl Vorwerck, Paul Karrenbrock, Walter Heinrich, and Hans Reupke; the last was both a member of the executive committee of the RDI and a Nazi, who sought to bring the corporatist interests of the Nazis in line with those of heavy industry.[19] By 1930, therefore, German industrialists were proposing various corporatist schemes for economic order and constitutional revision, though these proposals were decidedly not the same as those emanating from the ranks of small producers organized in the *Mittelstand* organizations or from a number of Nazis. Rather, they were explicitly defined as measures that would defend the "free economy" from the interference of trade unions, political parties, the parliament, and an intrusive state.[20]

During the fall of 1932, when leading representatives of domestic heavy industry and the RDI were negotiating with the Nazis, including Hitler, to secure guarantees against excessive state intervention, they made their case in defense of employer "freedom" in terms of their desire for the "self-organization of the corporatively organized economy," and they were not without success.[21] In response, Hitler "pulped" the Nazi "immediate economic program," sidelined the protectionist small-producer (*mittelständisch*) corporatisms of Gottfried Feder and Otto Wilhelm Wagner, and appointed the industry-friendly Walther Funk, former chief editor of the *Berlin Exchange Newspaper,* to head the agency in charge of the "private economy" within the Nazi Party.[22] Moreover, after their "seizure" of power, Nazi and DAF leaders curbed their earlier calls for dirigiste "corporatist reconstruction" that would limit the prerogatives of employers during the summer and fall of 1933, in favor of a vague plan for economic reorganization that would leave autonomy and decision-making powers in the hands of German employers and would not dismantle the structures of private property, market relations, and the profit mechanism—a reorientation of Nazi corporatism that was brought about in discussions with and conciliatory responses to German industrialists and their representatives. These discussions were facilitated by the Institute for Corporatism (Institut für Ständewesen), founded by the Nazi industrialist Fritz Thyssen and the Nazi Josef Klein of I.G. Farben in May 1933; the Office for Corporatist Reconstruction (Amt für ständischen Aufbau) of the Nationalsozialistische Deutsche Arbeiterpartei (hereafter NSDAP), an office directed by Max Frauendorfer; and direct contacts between leaders of

German heavy industry and Nazi officials over the course of 1933 and 1934.[23] The gradual process by which industrialists negotiated with Nazi leaders from 1931 to 1933 therefore involved not only the growing willingness of Nazi leaders to modify their "anticapitalist" views and subordinate their "socialist" wing in order to accommodate leading industrialists from domestic heavy industry; it also involved the steady disqualification of certain corporatist projects in favor of others, especially versions favored by heavy industry—not, as many historians have long argued, the abandonment of "corporatism" altogether.[24]

Indeed, after the repudiation of anticapitalist elements within the Nazi Party and the turn away from the state protectionism and the general project of *ständische Aufbau* elaborated on behalf of the *Mittlestand* and peasantry in 1934, a new framework of corporatist relations between state and industry was established as part of the rearmament drive of the 1930s. This new system of Reich, industry, and business groups and an architecture of cartels, established from 1933 to 1936, was designed to involve German employers in the industrial expansion and military rearmament policies of the Nazi state—a process that involved hard negotiations between representatives of German industry and state and party officials, not state efforts at "dragooning private enterprise" into "serving" Nazi "desires."[25] Thus the integration of the RDI into the new Imperial Estate of German Industry (Reichsstand der deutschen Industrie) in July 1933 was the outcome of joint efforts between Nazi officials and industrialists within the RDI—namely, figures like Fritz Thyssen and Jakob Reichert of the VDESI. During the summer of 1933, the new economics minister, Director Kurt Schmitt of the Allianz insurance company, negotiated with Hitler important spaces of self-regulation and autonomy for heavy industry in the new structure of cartels, leaving little room for the protectionist corporatism of the *Mittelstand.*[26] Moreover, the subsequent reorganization of industrial organizations to include new committees and "rings," which took place under the direction of Albert Speer from December 1941 to May 1942, was also the outcome of joint efforts between representatives of German industry and state and party officials. Inspired by Vögler and overseen by Walther Schieber, a former manager for I.G. Farben and director of a fiber company in Thuringia during the 1930s, these agencies were chaired mostly by representatives from Germany's leading one hundred industrial firms under the Nazi (and now Speer's) slogan of "self-administration." As Adam Tooze compellingly demonstrates, these organizations evolved into a "complex structure of cooperation" that drew on the "enthusiasm" of German industrialists and their representatives, especially from the coal, steel, chemical, automotive, and aircraft industries, who actively facilitated German plans for rearmament and imperial expansion on the continent.[27]

In light of these connections and shared ideological commitments, we need to re-think historiographical interpretations of the "Nazi economy" as a "planned economy" imposed by autonomous state actors from above in ways that subverted the "normal" workings of capitalism or interpretations that stress Nazi efforts, dictated by state officials, to build a "nationalistic-statist economic order" within a broadly capitalist framework.[28] The general process of corporatist restructuring described here is probably best characterized in terms of increasing "governmentalization of the state" in the formation of a new regime of capital accumulation, rather than increasing state intervention into an autonomous economic realm during the Third Reich.[29] This involved the strategic recodification of existing domains and practices of economic (self-)governance—the redirection of the independent activity of industrial interest groups into the expanding and evolving corporatist structures of the Nazi state. These structures certainly became mandatory for all employers, but they were put in place and managed by political-economic leaders drawn from the ranks of the Nazi movement and German industry. In these new corporatist configurations, they became part of a highly dynamic and restlessly expansionist regime of capital accumulation and mode of regulation. The latter regime constituted a repressively antiunion, militarized, and autarkic version of Fordism, which rested on a new balance of production and consumption imperatives; new regulatory mechanisms for pricing, credit, and industrial policy; enhanced schemes of rationalized mass production; and a new architecture of economic interest representation.[30]

This new regime did not satisfy the demands of all German employers across industrial sectors, but the lack of complete agreement among all businessmen and industrialists—a constitutive element of any capitalist system—should not obscure the important contributions of many leading German industrialists and their representatives to the collapse of the Weimar Republic and the rise of the NSDAP in 1932–33. These contributions are not best understood in terms of isolated tactical positions adopted by certain industrial leaders in the ebb and flow of political calculation from 1930 to 1933—not least because the connections between industrialists and the Nazi movement varied considerably during these years.[31] Rather, their involvement should be understood in terms of their relation to the corporatist political and ideological logic, anchored in the Papen-Hitler coalition, that finally destroyed the Republic. Moreover, the disagreements between and among German employers and Nazis should not obscure the role of industrialists, often as both businessmen and Nazis, in the policy-making processes of the Nazi state after 1933, including during the Second World War, as the total war economy drew on the technical expertise and resources of German industrialists in what was a corporatist structure of industrial production. Finally, recognition of the important changes in economic developments during the 1930s and 1940s and disagreements between and among German industrialists and Nazi political leaders should not overlook the capitalist-technocratic orientation of their authoritarian-corporatist projects—corporatisms that were not the same as prewar corporatism in the Saar but whose central terms and propositions and whose institutional forms were certainly anticipated by the latter.

This study focuses on the prewar era in the industrial Saar and cannot fully trace these complex genealogies into the 1920s and 1930s, but it can offer a glimpse of the fascist potential embedded in this ideological transformation and of the complex connections between individuals and corporatist ideology, by examining briefly the subsequent biographies of the two most important industrial representatives from the Saar: Max Schlenker and Herrmann Röchling. Schlenker remained in charge of the main Saar business organizations during the First World War and, in this capacity, was instrumental in orchestrating financial and organizational support from Saar heavy industry—along with other important industrial interests from the Ruhr and organized in the CVDI—for the new German Fatherland Party created in 1917.[32] In 1925, he moved to the Ruhr and became the principal managing director of the Langnamverein, and he served as a spokesman and leading "organic intellectual" from the ranks of Ruhr heavy industry during the Weimar era. From his position in the Ruhr, Schlenker served the cause of employer-friendly rationalization and the articulation of new ideologies of workplace management and company social policy, in his capacity as a member of the managing committee of Dinta.[33] But his most important ideological and organizational contributions to the politics of industrial interest groups during the Weimar years were related to his role as manager of the Langnamverein.

In this capacity, Schlenker was one of the leading advocates of a corporatist restructuring of the political order as a means of destroying the parliamentary and social welfare systems of the Weimar Republic. In 1925, along with Paul Reusch, general director of the Gutehoffnungshütte and chairman of the Langnamverein, he authored a set of political demands designed to counter the influence of those industrialists, such as Paul Silverberg and his allies within the RDI, who sought accommodation with the SPD and the institutions of the Weimar Republic. The document amounted to a general attack on the foundations of the Weimar settlement, with its demands for constitutional revision that would curb the spending or the "extraordinary luxury of parliament," dismantle the compulsory arbitration system for labor disputes, end price supports, rescind legislation against cartels, and disallow foreign "dumping" and cheap export credits. Schlenker envisioned a kind of "economic dictatorship"—enabled by article 48 of the Weimar constitution—that would confer on the government "especially strong administrative authority" and furnish the Reich finance minister with a "right to veto in cases involving decisions coming from all other departments or from the parliament that impose financial burdens on the budget."[34] These demands were never implemented, and the designs of Schlenker and other Ruhr industrialists were thwarted during the period of relative stabilization from 1926 to 1929, when the Silverberg line prevailed in the RDI. In November 1928, however, when the Northwest Group of German Iron and Steel Industrialists locked 230,000 steelworkers out their factories in an effort to challenge the decisions of state arbitrators during a wage dispute in the Ruhr, Schlenker once again played an important and effective role in the "public relations work" of domestic heavy industry. The action, which targeted the wage negotiation system in general as a foundation of the Weimar Republic, was an effort to block unwanted involvement of state agencies and political parties in "the free economy," in Paul Reusch's words; and the final compromise decision reached by SPD interior minister Carl Severing was condemned by Schlenker as a step in the direction of socialism.[35] The following year, Schlenker became involved in the nationalist agitation over the Young Plan, with its coordination among a broad range of organizations, including the German Nationalist People's Party, the Nazi Party, and the major right-wing veterans' association, the Stahlhelm. In this first mobilization of an emergent "nationalist opposition," Schlenker helped to steer leaders of the Langnamverein away from rejection of the new plan and toward a moderate line accepting its payment terms.[36]

During the summer of 1930, however, in the midst of economic depression, Schlenker and other representatives from German heavy industry began more aggressively to mount their corporatist "political offensive" against the presidential regime of Heinrich Brüning.[37] At the April 1930 meeting of the Langnamverein, Schlenker called on "the German employer" to become politically active in order "to secure his deserved position at the helm of the state." He then published an article in *Stahl und Eisen* arguing for the need to learn from Mussolini's corporative economic experiments in Italy, since the Reich Economic Council in Germany had not been able to contain the power of the parliament, the party fractions, and the parties. The latter forces, he maintained, had prevented an "economic policy of greater objectivity [*Sachlichkeit*]."[38] Later that year, in response to the demands for "economic democracy" emanating from the SPD, Schlenker called for the formation of a unified "front" among employer-parliamentarians, the creation of "political directors" in all large-scale industrial concerns, and a coalition among old and new political parties and even sections of youth, who understood the "significance of free economic production to the foundation of a people and state."[39] In 1931, Schlenker stepped up his political activity by publishing articles and undertaking a public speaking tour, mostly to military schools and meetings of officers, in order to describe the "objective economic situation" and make the case for political change.[40] In October 1931, he participated in the meeting of the Harzburg Front, an important alliance of the leading elements of the nationalist opposition to the Brüning cabinet, led by the NSDAP and the German Nationalist People's Party and including key representatives from German heavy industry. At this time, Schlenker emerged as a leading figure in attempts to forge a new authoritarian government of national opposition, under the leadership of Franz von Papen, who shared the corporatist views of many German industrialists, but with Nazi participation. The aim of Schlenker and other Ruhr industrialists was to appeal to what they viewed as the "bourgeois-conservative" wing of the Nazi Party, which they hoped would provide the necessary mass support for their own constitutional and economic reforms.[41]

In the fall of 1932, Schlenker participated in discussions of leading industrialists (including the October meeting in the Club of Berlin) that were designed to frustrate a possible "black-brown bloc" between the Nazis and the Center Party in upcoming elections in November and to support the Papen regime and "constitutional reform."[42] Like other employers and their representatives at this stage, Schlenker openly favored a corporatist extraconstitutional solution to what he perceived to be the problem of an endangered "free economy." In an essay that combined capitalist appeals to the rights of individual employers, criticism of trade unions and a "socialistic" state, and corporatist demands for constitutional revision, Schlenker called for the formation of a dictatorial regime in order to bring about a "healthy economy in a strong state."

> The parties here have for a long time no longer served as the expression [*Sprachrohre*] of political will in relation to the government and are no longer indispensable watchdogs [*Überwachungsorgane*] in the service of healthy governmental leadership. . . . Now it is necessary to establish guarantees that these conditions of the past never return. . . . We must never allow governments that are dependent on the Reichstag or political parties, as they have been in the past years, to return under any circumstances. The sense and purpose of constitutional reform is in the end the guarantee of a powerful government independent of the parties.[43]

Like the corporatist commitments of other German industrialists and their representatives, this vision combined demands for political order based on representation by economic "estate" with capitalist notions of employer prerogative, individual "performance" at work, the "factory community," competitive production, and economic growth. It sought a new regime of accumulation, which would destroy the power of the trade unions and the German Left, curb legislation that reduced profits, and abolish the industrial policy and social welfare guarantees of the Weimar Republic—all of which were deemed costly to employers and unnecessary interference in the "free economy."

Despite his efforts to destroy the Weimar Republic and to bring about an authoritarian order by means of a government of national concentration involving the National Socialists, the Nazis ultimately forced Schlenker out of his position as manager of the Langnamverein, owing to his preference for Papen over Hitler as chancellor.[44] This had little to do, however, with fundamental ideological differences in relation to economic and political order. Much of what Schlenker had long espoused—the destruction of parliamentary democracy, the elimination of the trade unions and collective bargaining, the reconfiguration of industrial class relations in accordance with the principles of "performance" and the "factory community," and the direct participation of economic interest organizations in the agencies of the state—was brought about, if in various ways, during the Third Reich. Schlenker's wider ideological and political contributions matter most in this context, not the extent of his personal involvement with or direct funding of the Nazi Party or Hitler himself. Schlenker's preference for a Papen-led dictatorial regime with Nazi support and his subsequent removal from leadership in the Langnamverein do not invalidate what became the fascist elements of his economic and political vision.

The postwar biography of Herrmann Röchling illustrates this point about the multiple articulations of ideological discourse and the crucial ideological affinities between powerful industrialists, their representatives, and the Nazis during the 1930s and 1940s. Whereas Schlenker was removed from his position by the Nazis in 1933, Röchling proved capable of being a successful Saar industrialist, an effective industry spokesman, and a prominent Nazi. During the interwar period in the Saar, his economic and political activities both prepared the way for the success of the NSDAP during the Saar plebiscite of 1935 and facilitated many of the economic, colonial, and terroristic ambitions of the Nazi regime after 1935. Before the war, Röchling, the leading proponent of the yellow union movement and the "works community," was drawn to the racist ideology of the radical Right, especially the Pan-Germanism of Heinrich Class. Like Schlenker, the Röchlings, along with Theodor Müller of the Stumm concerns and Edmund Weisdorff of the Burbach steelworks, supported the annexationist German Fatherland Party, the right-wing formation anchored in the political-ideological logic of the prewar Cartel of Productive Estates.[45]

After the German defeat, the separation of the Saar region from Germany, and the onset of French administration of the Saar under a League of Nations mandate, Röchling became the leader of the German-Saarland People's Party (Deutsch-Saarländische Volkspartei, hereafter DSVP) and the leading nationalist publicist in the region, a one-man "institution" who set up "Saar offices," with his social secretary Karl Rupp, in Völklingen and in other cities in Germany, in order to contest French authority and influence in the region.[46] Röchling and other members of the AVSI sponsored the Saar's own center for industrial sociology and "human management," the Institute for Labor Science (Anstalt für Arbeitskunde), created in 1927, which was located in Burbach and supported by all leading firms of Saar private heavy industry.[47] In addition, in 1928, he was one of the organizers of the League for the Renewal of the Reich (Bund zur Erneuerung des Reiches), an avowedly antidemocratic corporatist political organization created by Reusch and other industrialists but also including bankers and landowners and chaired by the former chancellor and German People's Party politician Hans Luther. The league called for "constitutional reform" that would nullify the independence of the Prussian government, install an authoritarian executive for Germany, and create a new legislative body or "Reich council" composed of representatives from occupational groups and experts.[48] Like the other prominent members of the league, which also included representatives from the Stumm steelworks in Neunkirchen and the Vopelius glassworks in Sulzbach, Röchling opposed the "socialism" and radicalism of the Nazis during this period but did so from an antidemocratic and antiparliamentary corporatist perspective that favored a government of national opposition led by a strong leader.[49]

This changed with the appointment of Hitler as chancellor in 1933 and the campaigns for the Saar plebiscite of 1935. Röchling was the principal local architect of the German Front, the multiparty nationalist coalition and propaganda organization formed in May 1933 for the purpose of mobilizing *Saarländer* in advance of the plebiscite. He contacted Hitler in order to begin discussions about establishing the front; took part in its organizational meetings of 1933 and 1934, which brought the Nazis into coalition with the DSVP, the Center, and the Economic Party of the Saar; and participated in the radio, film, and public speaking activities associated with the front's nationalist mobilization.[50] The front was a mass-based organization, which, like the Nazi Party, was organized from the cell and block (neighborhood) levels upward. It blanketed the region with nationalist propaganda from 1933 to 1935, saturating and dominating the local-regional "public sphere" with its messages and demands. This involved not only exhortations calling for the return of the Saar to Germany in public radio announcements and newspaper articles but also mass demonstrations, threats, and coercive measures that targeted the neighborhoods and factories. The front's "opening up and domination of private everyday life" included demonstrations and propaganda activities at the main iron and steel factories, including the Röchling concerns in Völkingen, where company celebrations, associations, and facilities were mobilized on behalf of the nationalist cause and where workers were forced to participate in Nazi symbolic practices, ranging from flying the Nazi flag to using the "Hitler greeting."[51] By May 1934, Röchling had obviously become comfortable with Nazi radicalism and "brown terror." In an article for the journal *Westland,* he openly threatened the antifascist opposition to reincorporation of the Saar.

> There are some *Saarländer* who have decided in favor of French citizenship. We have marked their names in the "official journal," and we

> have taken notice of them. But there might also be the other sort whose names we do not know. We think that it would be good for them if they were on the other side of the border when the plebiscite is over.[52]

Moreover, Röchling had given up his earlier misgivings about the economic policies of Hitler and the NSDAP and was insisting that Hitler and the Nazis promised not only national renewal but the "reconstruction of the German economy."[53]

After the plebiscite in January 1935 and the reincorporation of the Saar into Germany, Röchling officially became a member of the Nazi Party—at a time when the "old fighters" of the party were being pushed aside—and its most important industrialist in the Saar, where he took part in efforts to forge the "works community" by means of ideological indoctrination, coercion, and terror.[54] This involved the implementation of the new labor laws and institutions, along with their accompanying terminology, which Röchling had staunchly supported; the incorporation of DAF activities and social policy programs; and seeking out enemies of the new order among the labor force in order to secure a "cleansed works community."[55] The labor policies designed to consolidate the "works community" and secure greater workplace "performance" among workers radically expanded in the context of war after 1939, when Röchling and his officials availed themselves of the "work reeducation camp" created in Hinzert, a few kilometers north of the Saar border in the Hunsrück, in October 1939.[56] Initially inspired by Fritz Todt, general inspector of the Nazi labor and construction office called Organization Todt, the camp at Hinzert became the model for an expanding network of "reeducation" centers, subsequently initiated by German employers and designed to incarcerate workers, under brutal conditions, for the "crimes" of absenteeism, slowdowns, lack of productivity, drinking on the job, insubordination, expressions of opposition to the Nazi regime, and actions deemed detrimental to the successful "prosecution of the war."[57] This effort to terrorize workers at the Röchling steelworks reached its peak when the firm set up its own "factory summary court" (*Betriebliche Schnellgericht*) in April 1943. The court, which was run by company officials, state "trustees of labor," and representatives from the DAF, would "sentence" workers for offenses related to "idleness at work" (*Arbeitsbummelei*), after which they were incarcerated in a factory-owned camp in Etzenhofen. Targeting German workers and the ever-growing numbers of foreign workers, mostly Eastern European slave laborers, it represented a privatized form of managerial terror, which, according to Röchling, was designed to raise the level of workplace discipline without losing workers to the camps of the gestapo.[58]

Finally, Herrmann Röchling ultimately became "one of the most powerful industrialists" in the corporative structure and war machine of the Nazi empire.[59] A long-standing advocate of imperial expansion, racist and anti-Semitic nationalism, and German continental hegemony, Röchling was a close ally of Hermann Göring, Reich marshal and head of the Four-Year Plan Office after 1936.[60] These commitments and connections led to his appointment as chairman of the Reich Group Industry, District Group Saarland-Palatinate (Bezirksgruppe Saarland-Pfalz der Reichsgruppe Industrie) and to his later appointment as chairman of the Reich Iron Association (Reichsvereiningung Eisen), an organization of all iron and steel firms in Germany. Created in May 1942, the Reich Iron Association was designed as a "national peak cartel" for the purpose of organizing iron and steel production for the regime's war and expanding continental hegemony and "Greater German Empire." It took the form of a corporatist organization responsible for rationalizing and increasing iron and steel production, promoting technological advancements in the iron and steel industry, coordinating the allocation of raw materials and the import and export the industry's products, overseeing existing cartels, setting prices and regulating disputes among producers, and making determinations over the profitability and viability of iron and steel firms. These vital functions, according to Röchling, were not to serve an anticapitalist system of centralized state planning but, rather, were designed to maintain German capitalism or the "private economy."[61]

In this capacity, Röchling influenced labor and economic policy throughout the iron and steel industry of the Third Reich during the war years. Röchling was thus responsible for ensuring the exercise of terror at the workplace in the iron and steel industry, including issuing orders to the member firms of the Reich Iron Association to extend punishment of the "transgressions" of steelworkers on the shop floor "all the way to the concentration camp if necessary";[62] and as chairman, he helped to organize the economic empire of the Nazi regime as it extended into newly occupied territories. In July 1942, Röchling was appointed Reich representative for iron and steel in the occupied territories and took charge of matters related to economic policy and industrial ownership and concentration in the newly conquered regions of Europe. His efforts on behalf of the expanding German empire, especially in the occupied "western regions," served his own economic interests (for Röchling had long sought access to the ore fields of Lorraine) and the wider economic interests of Nazi regime.[63] Consequently, Röchling was put on trial in Rastatt in the French occupation zone in February 1948 and sentenced by an international tribunal to ten years in prison for his activities on behalf of the Nazi regime, especially as "promoter and beneficiary" of "slave labor."[64]

There was nothing predetermined about Röchling's path from prewar leader of the Saar and national yellow union movement and representative for Saar heavy industry to Reich representative for iron and steel during the Third Reich; nor can the history of corporatism be traced in linear fashion from the decade before 1914 to the 1930s and 1940s, for there were important differences between the emergent corporatist political project of employers before the First World War and the officially sanctioned architecture of interest organizations, regulatory mechanisms, and violence of Nazi "corporatism" during the 1940s. Rather, these histories are best understood as complex genealogies—contingent and "heterogeneous" trajectories that shifted and evolved in relation to other "discourses, practices, and events" from 1900 to 1945.[65] Instead of a history of the continuity of industrial relations, employer politics, and the radical Right from 1900 to 1945, this genealogical perspective offers analysis of the conditions of emergence and formation of radical nationalist corporatism. It suggests that those conditions were produced in a set of wider social and political transformations after the 1890s: rapid economic and organizational concentration in German domestic heavy industry, the creation of new forms of state welfare and workplace regulation, the reconstitution of the forms and domains of *Öffentlichkeit,* and political struggles over factory paternalism itself, which eroded the foundations of the paternalist sociopolitical order in the Saar and other parts of Germany. In this context, employers and their representatives responded to the challenges of bourgeois liberals, Catholic reformers, Christian trade unionists, and Social Democrats by reorganizing and rearticulating the structures of work, economic interest organization, and sociopolitical order, in distinctively corporatist ways. This new ideological discourse, which incorporated technocratic and bioracial schemas of workplace management, aimed self-consciously to restore employer prerogative and the conditions of capitalist profitability in a "free economy," which was allegedly threatened by the rise of trade unions, the political Left, and parliamentary democracy. Max Schlenker carried this corporatist discourse over into the Weimar era and rearticulated it when he attacked the Weimar Republic on behalf of Ruhr heavy industry, though his tactical opposition to Hitler's chancellorship cost him his job and influence after 1933. In the case of Hermann Röchling, however, long-espoused capitalist labor policies and economic practices, corporatist forms of industrial and social organization, biological racism, and imperialist ideologies anticipated and were effectively adapted to the industrial order and colonial empire of the Third Reich.

Notes

INTRODUCTION

1. "Verhandlungen des Vereins für Socialpolitik über das Arbeitsverhältnis in den privaten Riesenbetrieben," in *Schriften des Vereins für Socialpolitik, 116 Band, Verhandlungen der Generalversammlung in Mannheim, 25., 26., 27. und 28. September 1905* (Leipzig: Duncker & Humblot, 1905), 135–235.

2. Ibid., 214–15, 213.

3. Ibid., 157.

4. See especially Hans-Ulrich Wehler, *The German Empire, 1871–1918* (Leamington Spa: Berg Publishers, 1985).

5. Dirk Stegmann, "Zwischen Repression und Manipulation. Konservative Machteliten und Arbeiter- und Angestelltenbewegung 1910–1918. Ein Beitrag zur Vorgeschichte der DAP/NSDAP," *Archiv für Sozialgeschichte* 12 (1972): 358.

6. Wehler, "Wie 'bürgerlich' war das Deutsche Kaiserreich?" in *Aus der Geschichte lernen?* (Munich: C. H. Beck, 1988), 210; Wehler, *Deutsche Gesellschaftsgeschichte,* vol. 3, *Von der "Deutschen Doppelrevolution" bis zum Beginn des Ersten Weltkrieges 1849–1914* (Munich: C. H. Beck, 1995), 726–29; Gerhard A. Ritter and Klaus Tenfelde, *Arbeiter im Deutschen Kaiserreich 1871 bis 1914* (Bonn: J. H. W. Dietz, 1992), 409–13.

7. Jürgen Kocka, "Vorindustrielle Faktoren in der deutschen Industrialisierung. Industriebürokratie und 'neuer Mittelstand,'" in *Das kaiserliche Deutschland. Politik und Gesellschaft 1870–1918,* ed. Michael Stürmer (Düsseldorf: Droste Verlag, 1970), 265–86; Kocka, *Unternehmensverwaltung und Angestelltenschaft am Beispiel Siemens 1847–1914. Zum Verhältnis von Kapitalismus und Bürokratie in der deutschen Industrialisierung* (Stuttgart: Klett, 1969), 234, 310, 353; Hartmut Berghoff, "Unternehmenskulture und Herrschaftstechnik. Industrieller Paternalismus. Hohner von 1857 bis 1918," *Geschichte und Gesellschaft* 23 (1997): 167–204.

8. On antimodernism, see Fritz Stern, *The Politics of Cultural Despair* (Berkeley and Los Angeles: University of California Press, 1961), xvi–xvii, 58, 230–31, 258; and George Mosse, *The Crisis of German Ideology* (New York: Universal Library, 1964), 13ff., 282–91. For the small producers and domestic heavy industry, see Heinrich August Winkler, *Mittelstand, Demokratie und Nationalsozialismus. Die politische Entwicklung von Handwerk und Kleinhandel in der Weimarer Republik* (Cologne: Kiepenheuer & Witsch, 1972), 114; and Winkler, "Unternehmerver-

bände zwischen Ständeideologie und Nationalsozialismus," *Vierteljahrshefte für Zeitgeschichte* 17 (1969): 369. For an indispensable discussion of heavy industry, which provides the quote about "conservative feudal value orientations," see Dirk Stegmann, "Hugenberg Contra Stresemann. Die Politik der Industrieverbände am Ende des Kaiserreichs," *Vierteljahrshefte für Zeitgeschichte* 24 (1976): 336–37, 350; Stegmann, "Zwischen Repression und Manipulation," 354, 357; and Stegmann, *Die Erben Bismarcks. Parteien und Verbände in der Spätphase des Wilhelminischen Deutschlands. Sammlungspolitik 1898–1918* (Cologne and Berlin: Verlag Kiepenheuer & Witsch, 1970), 113–28.

9. Mommsen, "Sozialpolitik im Ruhrbergbau," in *Industrielles System und politische Entwicklung in der Weimarer Republik,* ed. Hans Mommsen, Dietmar Petzina, and Bernd Weisbrod (Düsseldorf: Droste Verlag, 1974), 307–8, 321; Winkler, "Unternehmerverbände zwischen Ständeideologie und Nationalsozialismus," 369–71.

10. See the classic critique in David Blackbourn and Geoff Eley, *The Peculiarities of German History: Bourgeois Society and Politics in Nineteenth-Century Germany* (Oxford: Oxford University Press, 1984).

11. Eley, "Capitalism and the Wilhelmine State: Industrial Growth and Political Backwardness, 1890–1918," in *From Unification to Nazism: Reinterpreting the German Past* (Boston: Allen & Unwin, 1986), 42–58; Crew, *Town in the Ruhr: A Social History of Bochum, 1860–1914* (New York: Columbia University Press, 1979), 157.

12. Eley, "Capitalism and the Wilhelmine State," 47, 50.

13. Dick Geary, "The Industrial Bourgeoisie and Labour Relations in Germany, 1871–1933," in *The German Bourgeoisie: Essays on the Social History of the German Middle Class from the Eighteenth to the Early Twentieth Century,* ed. David Blackbourn and Richard Evans (London and New York: Routledge, 1991), 140–61. See also Bernd Weisbrod, "Entrepreneurial Politics and Industrial Relations in Mining in the Ruhr Region: From Managerial Absolutism to Co-determination," in *Workers, Owners, and Politics in Coal Mining: An International Comparison of Industrial Relations,* ed. Gerald D. Feldman and Klaus Tenfelde (New York, Oxford, and Munich: Berg, 1990), 141–44. For other studies of the imperial era, see Patrick Fridenson, "Herrschaft im Wirtschaftsunternehmen. Deutschland und Frankreich 1880–1914," in *Bürgertum im 19. Jahrhundert. Deutschland im europäischen Vergleich,* vol. 2, ed. Jürgen Kocka (Munich: Deutscher Taschenbuch Verlag, 1988), 65–91; Ulrich Zumdick, *Hüttenarbeiter im Ruhrgebiet. Die Belegschaft der Phoenix-Hütte in Duisburg-Laar 1853–1914* (Stuttgart: Klett-Cotta, 1990); Elaine Golvka Spencer, *Management and Labor in Imperial Germany: Ruhr Industrialists as Employers, 1896–1914* (New Brunswick, NJ: Rutgers University Press, 1984), 71–89; and Thomas Welskopp, "Betriebliche Sozialpolitik im 19. und frühen 20. Jahrhundert," *Archiv für Sozialgeschichte* 34 (1994): 333–73.

14. Nolan, *Visions of Modernity: American Business and the Modernization of Germany* (Oxford: Oxford University Press, 1994), 180, 190.

15. I have developed this definition of the modern more fully in my article "Reconsidering the Modernity Paradigm: Reform Movements, the Social, and the State in Wilhelmine Germany," *Social History* 31 (2006): 430–34.

16. Geary, "Industrial Bourgeoisie," 152; Geary, "Employers, Workers, and the Collapse of the Weimar Republic," in *Weimar: Why Did German Democracy Fail?* ed. Ian Kershaw (London: Weidenfeld and Nicolson, 1990), 105; Thomas Welskopp, *Arbeit und Macht im Hüttenwerk. Arbeits- und industrielle Beziehungen in der deutschen und amerikanischen Eisen- und Stahlindustrie von den 1860er bis zu den 1930er Jahren* (Bonn: J. H. W. Dietz, 1994), 670–94; Nolan, *Visions of Modernity,* especially 232–35; Detlev Peukert, *Max Webers Diagnose der Moderne* (Göttingen: Vandenhoeck & Ruprecht, 1989), 81–82.

17. Mason, "The Primacy of Politics: Politics and Economics in National Socialist Germany," in *The Nature of Fascism,* ed. Stuart J. Woolf (London: Weidenfeld and Nicolson, 1968), 165–95. The essay was originally published in German as "Der Primat der Politik. Politik und Wirtschaft im Nationalsozialismus," *Das Argument* 41 (1966): 73–94.

18. Henry Turner, *German Big Business and the Rise of Hitler* (Oxford: Oxford University Press, 1985), 332; Peter Hayes, *Industry and Ideology: IG Farben in the Nazi Era* (Cambridge: Cambridge University Press, 1987), 85; Peter Hayes, *From Cooperation to Complicity: Degussa in the Third Reich* (Cambridge: Cambridge University Press, 2004); Gerald D. Feldman, *Allianz and the German Insurance Business, 1933–1945* (Cambridge: Cambridge University Press, 2001). For a perhaps more nuanced set of arguments, see Francis R. Nicosia and Jonathan Huener, eds., *Business and Industry in Nazi Germany* (New York and Oxford: Berghahn Books, 2004).

19. R. J. Overy, *War and Economy in the Third Reich* (Oxford: Oxford University Press, 1994); Paul Erker, *Industrie-Eliten in der NS-Zeit. Anpassungsbereitschaft und Eigeninteresse von Unternehmen in der Rüstungs- und Kriegswirtschaft, 1936–1945* (Passau: Wissenschaftsverlag Rothe, 1993); Lothar Gall and Manfred Pohl, eds., *Unternehmen im Nationalsozialismus* (Munich: C. H. Beck, 1998); Peter Hayes, "Polycracy and Policy in the Third Reich: The Case of the Economy," in *Reevaluating the Third Reich,* ed. Thomas Childers and Jane Caplan (New York: Holmes & Meier, 1993), 190–210; Gerhard Th. Mollin, *Montankonzerne und "Drittes Reich." Der Gegensatz zwischen Monopolindustrie und Befehlswirtschaft in der deutschen Rüstung und Expansion 1936–1944* (Göttingen: Vandenhoeck & Ruprecht, 1988)

20. Dieter Gosewinkel, "Wirtschaftskontrolle und Recht in der nationalsozialistischen Diktatur. Forschung und Forschungsperspektiven," in *Wirtschaftskontrolle und Recht in der nationalsozialitischen Diktatur,* ed. Gosewinkel (Frankfurt am Main: Vittorio Klostermann, 2005), xxi; Adam Tooze, *Wages of Destruction: The Making and Breaking of the Nazi Economy* (London: Penguin, 2006), an important new study; Thad Michael Allen, *The Business of Genocide: The SS, Slave Labor, and the Concentration Camps* (Chapel Hill: University of North Carolina Press, 2002). See, above all, Götz Aly and Susanne Heim, *Vordenker der Vernichtung. Auschwitz und die deutschen Pläne für eine neue europäische Ordnung* (Frankfurt am Main: Fischer Taschenbuchverlag, 1993), a pathbreaking work; and Christian Gerlach, *Kalkulierte Morde. Die deutsche Wirtschafts- und Vernichtungspolitik in Weißrußland 1941 bis 1944* (Hamburg: Hamburger Edition, 1999).

21. Geoff Eley, *A Crooked Line: From Cultural History to the History of Society*

(Ann Arbor: University of Michigan Press, 2005), 11. For other excellent recent guides to these debates, see William H. Sewell Jr., *Logics of History: Social Theory and Social Transformation* (Chicago: University of Chicago Press, 2005), especially 1–21, 22–80; Miguel A. Cabrera, *Postsocial History: An Introduction* (Lanham, MD: Lexington Books, 2004); and Geoff Eley and Keith Nield, *The Future of Class in History: What's Left of the Social?* (Ann Arbor: University of Michigan Press, 2007).

22. Alf Lüdtke, "Einleitung. Was ist and wer treibt Alltagsgeschichte?" in *Alltagsgeschichte. Zur Rekonstruction historischer Erfahrungen und Lebensweisen,* ed. Lüdtke (Frankfurt am Main: Campus Verlag, 1989), 9–47; Kathleen Canning, "Gender and the Politics of Class Formation: Rethinking German Labor History," *American Historical Review* 97 (1992): 736–68.

23. Thomas Welskopp, "Klasse als Befindlichkeit? Vergleichende Arbeitergeschichte vor der kulturhistorischen Herausforderung," *Archiv für Sozialgeschichte* 38 (1998): 301–36; Gary Herrigel, *Industrial Constructions: The Sources of German Industrial Power* (Cambridge: Cambridge University Press, 1996); Richard Bernacki, *The Fabrication of Labor: Germany and Britain, 1640–1914* (Berkeley and Los Angeles: University of California Press, 1995); Thomas Welskopp, "Der Betrieb als soziales Handlungsfeld. Neuere Forschungssätze in der Industire- und Arbeitergeschichte," *Geschichte und Gesellschaft* 22 (1996): 118–42.

24. Margaret Somers, "Narrativity, Narrative Identity, and Social Action: Rethinking English Working-Class Formation," *Social Science History* 16 (1992): 609–11; George Steinmetz, "German Exceptionalism and the Origins of Nazism: The Career of a Concept," in *Stalinism and Nazism: Dictatorships in Comparison,* ed. Ian Kershaw and Moshe Lewin (Cambridge: Cambridge University Press, 1997), 268–69.

25. The critical starting points here are Raymond Williams, *Marxism and Literature* (Oxford: Oxford University Press, 1977), especially 21–44, 55–71, 165–72; V. N. Vološinov, *Marxism and the Philosophy of Language* (Cambridge, MA: Harvard University Press, 1986); Stuart Hall, "The Rediscovery of Ideology: Return of the Repressed in Media Studies," in *Culture, Society, and the Media,* ed. Michael Gurevitch, Tony Bennett, James Curran, and Janet Woollacott (London and New York: Routledge, 1982), 56–90; Hall, "The Problem of Ideology: Marxism without Guarantees," in *Stuart Hall: Critical Dialogues in Cultural Studies,* ed. David Morley and Kuan-Hsing Chen (London and New York: Routledge, 1996), 25–46.

26. Louis Althusser, *Lenin and Philosophy, and Other Essays* (London: NLB, 1971), 121–73; Hall, "Rediscovery of 'Ideology,'" 79.

27. Mikhail Bakhtin, *The Dialogic Imagination: Four Essays* (Austin: University of Texas Press, 1981), 270; Stuart Hall, "For Allon White: Metaphors of Transformation," in Morley and Chen, *Stuart Hall,* 296. See also Richard Terdiman, *Discourse/Counter-Discourse: The Theory and Practice of Symbolic Resistance in Nineteenth-Century France* (Ithaca, NY: Cornell University Press, 1985), 25–81.

28. On articulation, see Ernesto Laclau, *Politics and Ideology in Marxist Theory* (London: Verso, 1977), 81–142, especially 100–111; Stuart Hall, "On Postmodernism and Articulation: An Interview with Stuart Hall," in Morley and Chen, *Stuart Hall,* 131–50.

29. Margaret Somers, "Workers of the World, Compare!" *Contemporary Sociology* 18, no. 3 (1989): 329; Somers, "Class Formation and Capitalism. A Second Look at a Classic," *Archives européennes de sociologie* 37 (1996): 180–202. See also the critique in Eley and Nield, *Future of Class,* 101, 130–32.

30. Scott, *Gender and the Politics of History,* 56, 89; Scott, "The Meaning of Experience," *Critical Inquiry* 17 (1991): 773–97; Scott, "The 'Class' We Have Lost," *International Labor and Working-Class History* 57 (2000): 71. See also Gareth Stedman Jones, "The Determinist Fix: Some Obstacles to the Further Development of the Linguistic Approach to History in the 1990s," *History Workshop Journal* 42 (Autumn 1996): 19–35; Patrick Joyce, "The End of Social History?" *Social History* 20 (1995): 73–91; Patrick Joyce, *Democratic Subjects: The Self and the Social in Nineteenth-Century England* (Cambridge: Cambridge University Press, 1994); and Joan Scott, *Only Paradoxes to Offer: French Feminists and the Rights of Man* (Cambridge, MA: Harvard University Press, 1996).

31. This perspective agrees with Sewell's in *Logics of History* (318–72), a tour de force of social and cultural-theoretical revision.

32. Hall, "Postmodernism and Articulation," 141–42; Hall, "Problem of Ideology," 40.

33. Williams, *Marxism and Literature,* 39.

34. Bakhtin, *Dialogic Imagination,* 259; Peter Stallybrass and Allon White, *The Politics and Poetics of Transgression* (Ithaca, NY: Cornell University Press, 1986), 80–84, 194; Kathleen Canning, *Languages of Labor and Gender: Female Factory Work in Germany, 1850–1914* (Ithaca, NY: Cornell University Press, 1996), 10–11.

35. Eley, "Nations, Publics, and Political Cultures: Placing Habermas in the Nineteenth Century," in *Habermas and the Public Sphere,* ed. Craig Calhoun (Cambridge, MA: MIT Press, 1992), 306. See Habermas's classic *Strukturwandel der Öffentlichkeit. Untersuchungen zu einer Kategorie der bürgerlichen Gesellschaft* (Frankfurt am Main: Suhrkamp, 1990), 69ff., 226, 295.

36. Madeleine Hurd, *Public Spheres, Public Mores, and Democracy: Hamburg and Stockholm, 1870–1914* (Ann Arbor: University of Michigan Press, 2000), especially 265–66, 271; Ute Daniel, *Arbeiterfrauen in der Kriegsgesellschaft. Beruf, Familie und Politik im Ersten Weltkrieg* (Göttingen: Vandenhoeck & Ruprecht, 1989), 233–49; Belinda Davis, "Reconsidering Habermas, Gender, and the Public Sphere: The Case of Wilhelmine Germany," in *Society, Culture, and the State in Germany, 1870–1930,* ed. Geoff Eley (Ann Arbor: University of Michigan Press, 1996), 397–426. See also David Blackbourn's seminal analysis of the rise of a bourgeois public sphere in nineteenth-century Germany in Blackbourn and Eley, *Peculiarities of German History,* 192–205.

37. Miriam Hansen, "Early Silent Cinema: Whose Public Sphere?" *New German Critique* 29 (1983): 156. See also Oskar Negt and Alexander Kluge, *Öfftentlichkeit und Erfahrung. Zur Organisationsanalyse von bürgerlicher und proletarischer Öffentlichkeit* (Frankfurt am Main: Suhrkamp, 1972); Nancy Fraser, "Rethinking the Public Sphere: A Contribution to the Critique of Actually Existing Democracy," in *Justice Interruptus: Critical Reflections on the "Postsocialist" Condition* (New York and London: Routledge, 1999), 69–98; and Mike Hill and Warren Sontag, eds., *Masses, Classes, and the Public Sphere* (London: Verso, 2000).

38. Harold Mah, "Phantasies of the Public Sphere: Rethinking the Habermas of Historians," *Journal of Modern History* 72 (2000): 166–67.

39. Michael Warner, *Publics and Counterpublics* (New York: Zone Books, 2002), 95; Stallybrass and White, *Politics and Poetics,* 80–124, 199. The three-part definition of *Öffentlichkeit* is taken from Miriam Hansen's discussion of its meanings in the German context. See her foreword to Oskar Negt and Alexander Kluge, *Public Sphere and Experience: Toward an Analysis of the Bourgeois and Proletarian Public Sphere* (Minneapolis: University of Minnesota Press, 1993), ix.

40. Habermas, *Strukturwandel,* 84.

41. Eley, "German History and the Contradictions of Modernity: The Bourgeoisie, the State, and the Mastery of Reform," in Eley, *Society, Culture, and the State,* 94–95; Blackbourn and Eley, *Peculiarities of German History,* 96–97.

42. Steinmetz, *Regulating the Social: The Welfare State and Local Politics in Imperial Germany* (Princeton, NJ: Princeton University Press, 1993), 73–107; Canning, *Languages of Labor.*

43. Nicos Poulantzas, *State, Power, Socialism* (London: Verso, 1978), 129, 136, 58, 64–5.

44. Mitchell, "Society, Economy, and the State Effect," in *State/Culture: State-Formation after the Cultural Turn* (Ithaca: Cornell University Press, 1999), ed. George Steinmetz, 77, 83, 89; Mitchell, "The Limits of the State: Beyond Statist Approaches and Their Critics," *American Political Science Review* 85, no. 1 (1991): 77–96.

45. See the classic early statement about the role of the state in the creation of capitalism in Karl Polanyi, *The Great Transformation: The Political and Economic Origins of Our Time* (Boston: Beacon Press, 1957), especially 139ff. See also the important work in regulation theory, which is best approached through Bob Jessop's "Regulation Theories in Retrospect and Prospect," *Economy and Society* 19 (1990): 153–216. Herrigel, *Industrial Constructions,* is now the most important study of this kind of the relationship between state and economy in the field of German history.

46. Erich Müller, *Der Steinkohlenbergbau des Preußischen Staates in der Umgebung von Saarbrücken,* part 6, *Die Entwickelung der Arbeiterverhältnisse auf den staatlichen Steinkohlenbergwerken vom Jahre 1816 bis zum Jahre 1903* (Berlin: Verlagsbuchhandlung von Julius Springer, 1904), 40ff.; Klaus-Michael Mallmann and Horst Steffens, *Lohn der Mühen. Geschichte der Bergarbeiter an der Saar* (Munich: C. H. Beck, 1989), 42–54.

47. *Jahresbericht der Handelskammer zu Saarbrücken für 1910* (Saarbrücken: Buchdruckerei der Neuen Saarbrücker Zeitung, 1911), II.5, II.7; Horst Steffens, "Auf 'Jahrhunderte hinaus eine blühende Zukunft.' Der Staatsbergbau," in *Industriekultur an der Saar. Leben und Arbeit in einer Industrieregion 1840–1914,* ed. Richard van Dülmen (Munich: C. H. Beck, 1989), 51, 54.

48. Johannes Schmitt, "Agrarische Krise und industrieller Aufbruch. Die Sozialtopographie des Saarraums," in van Düllmen, *Industriekultur an der Saar,* 24–25.

49. Anton Haßlacher, *Das Indusriegebiet an der Saar und seine hauptsächlichsten Industriezweige* (Saarbrücken: Pechner'sche Buchdruckerei, 1912), 117.

50. Ibid., 125; Wilhelm Born, *Die wirtschaftliche Entwicklung der Saar-Grosseisenindustrie seit der Mitte des 19. Jahrhunderts* (Berlin: Deutscher Provinzverlag, 1919), 129ff.

51. Haßlacher, *Das Indusriegebiet an der Saar,* 151–54, 159–60; Harald Glaser, "Thomasstahl und Tafelglas. Die Eisen- und Glashütten zwischen 1840 und 1918," in van Dülmen, *Industriekultur an der Saar,* 69.

52. Wolfgang Laufer, "Bevölkerungs- und siedlungsgeschichtliche Aspekte der Industrialisierung an der Saar," *Zeitschrift für die Geschichte der Saargegend* 29 (1981): 122–64.

53. Alfred Ruppersberg, *Geschichte der Stadt Saarbrücken und St. Johann von 1815 bis 1909, der Stadt Malstatt-Burbach und der verinigten Stadt Saarbrücken bis zum Jahre 1914* (Saarbrücken: Verlag des Kreises und der Stadt Saarbrücken, 1914), 187, 221–71; Stefan Leiner, *Migration und Urbanisierung. Binnenwanderungsbewegungen, räumlicher und sozialer Wandel in den Indusriestädten des Saar-Lor-Lux-Raumes 1856–1910* (Saabrücken: Saarbrücker Drückerei und Verlag, 1994), 66–92; Klaus Fehn, "Das saarländische Arbeiterbauerntum im 19. und 20. Jahrhundert," in *Agrarisches Nebengewerbe und Formen der Reagrarisierung im Spätmittelalter und 19./20. Jahrhundert,* ed. Hermann Kellenbenz (Stuttgart: Gustav Fischer Verlag, 1975), 195–214.

CHAPTER I

1. BM Woytt (Gersweiler) to LR (Saarbrücken), 5 May 1890, in Landesarchiv Saarbrücken (hereafter LAS), Landratsamt Saarbrücken, S/3, unpaginated; LR (Saarbrücken) to OP (Trier), in Landeshauptarchiv Koblenz (hereafter LHAK), 442/4420, 40.

2. Donald Reid, "Industrial Paternalism: Discourse and Practice in Nineteenth-Century French Mining and Metallurgy," *Comparative Studies in Society and History* 27 (1985): 579–607.

3. See especially Kathleen Canning, "Gender and the Culture of Work: Ideology and Identity in the World behind the Mill Gate, 1890–1914," in *Elections, Mass Politics, and Social Change in Modern Germany: New Perspectives,* ed. Larry Eugene Jones and James Retallack (Cambridge: Cambridge University Press, 1992), 175–99; and the stimulating discussion in Joan Scott, *Gender and the Politics of History* (New York: Columbia University Press), 139–63, from which this general argument draws its inspiration.

4. Abraham Ascher, "Baron von Stumm, Advocate of Feudal Capitalism," *Journal of Central European Affairs* 22 (1962): 283, 275.

5. See, e.g., his speech to the Reichstag of the North German Confederation from October 1867, in Alexander Tille, ed., *Die Reden des Freiherrn Carl Ferdinand von Stumm-Halberg,* vol. 8 (Berlin: Otto Elsner, 1912), 21–24.

6. "Theorie und Praxis," *Saarbrücker Gewerbeblatt* (hereafter *SGB*), 27 December 1885, no. 52, 209; Stumm's comments during the Reichstag deliberations over the new Industrial Code Bill in 1890, in Alexander Tille, ed., *Die Reden des Freiherrn Carl Ferdinand von Stumm-Halberg,* vol. 11 (Berlin: Otto Elsner, 1914), 305.

7. These remarks were made during a bonus ceremony in Neunkirchen on 22 June 1895. See Alexander Tille, ed., *Die Reden des Freiherrn Carl Ferdinand von Stumm-Halberg,* vol. 12 (Berlin: Otto Elsner, 1915), 584–85.

8. Tille, *Die Reden,* 11:327, 329.

9. Ibid., 12:10.

10. Cited from Stumm's 21 June 1891 speech to his workers in Neunkirchen, in Tille, *Die Reden,* 12:561–62; Fritz Hellwig, *Carl Ferdinand von Stumm-Halberg 1836–1901* (Heidelberg-Saarbrücken: Westmark-Verlag, 1936), 298.

11. Tille, *Die Reden,* 12:19.

12. Alexander Tille, ed., *Die Reden des Freiherrn Carl Ferdinand von Stumm-Halberg,* vol. 10 (Berlin: Otto Elsner, 1913), 27.

13. Tille, *Die Reden,* 8:340; 12:309, 558.

14. Ibid., 12:558; Stumm's comments about English trade unions in Alexander Tille, ed., *Die Reden des Freiherrn Carl Ferdinand von Stumm-Halberg,* vol. 9 (Berlin: Otto Elsner, 1912), 435; Tille, *Die Reden,* 11:300, 12:586–87.

15. Tille, *Die Reden,* 10:28.

16. Ibid., 11:333–34. These comments were made before Stumm was forced to amend his marriage clause by the new industrial code of 1891. See also the reference to the "voluntary" nature of this kind of agreement in ibid., 11:329.

17. "Wohlfahrtseinrichtungen für Fabrikarbeiter," *SGB,* 27 April 1884, no. 17, 16–17.

18. Tille, *Die Reden,* 12:9.

19. Heinrich Junghahn, *Das Schlafhaus- und Einliegerwesen im Bezirk der Kgl. Bergwerksdirektion Saarbrücken* (Berlin: Wilhelm Ernst & Sohn, 1912), 8.

20. "Wohnhäuser für Arbeiter," *SGB,* 7 December 1890, no. 49, 199.

21. The quotes are from Tille, *Die Reden,* 12:587, 11:329, 12:564. See also ibid., 12:13, 11:333–34.

22. The rules are reprinted in ibid., 12:599–622. See also the factory notice dated 31 March 1892 in Neunkircher Eisenwerksarchiv (hereafter NEA).

23. The quotes are taken from the general *Arbeitsordnung* in Tille, *Die Reden,* 12: sections 14–36.

24. "Arbeits-Ordnung für die königlichen Steinkohlengruben bei Saarbrücken 1877," in LAS, 564/2343, 773.

25. Circular no. 23, 17 July 1884, and circular no. 45, 15 August 1887, in NEA.

26. See section 38 of the "Allgemeine Arbeitsordnung für das Neunkircher Eisenwerk 1895," in NEA.

27. For details, see Horst Steffens, *Autorität und Revolte. Alltagsleben und Streikverhalten der Bergarbeiter an der Saar im 19. Jahrhundert* (Weingarten: Drumlin, 1987), 204–7.

28. Steffens, *Autorität,* 205.

29. Cited in Klaus-Michael Mallmann and Horst Steffens, *Lohn der Mühen. Geschichte der Bergarbeiter an der Saar* (Munich: C. H. Beck, 1989), 35, 48. See also Junghahn, *Das Schlafhaus- und Einliegerwesen,* 8; Max Eichhorst, *Die Lage der Bergarbeiter im Saargebiet* (Eisleben: Ernst Schneider, 1901), 34; Erich Müller, *Der Steinkohlenbergbau des Preußischen Staates in der Umgebung von Saarbrücken,* part 6, *Die Entwickelung der Arbeiterverhältnisse auf den staatlichen Steinkohlenbergwerken vom Jahre 1816 bis zum Jahre 1903* (Berlin: Verlagsbuchhandlung von Julius Springer, 1904), 129; and Ernst Klein, "Bergfiskus und Kirche an der Saar im 19. Jahrhundert," *Zeitschrift für die Geschichte der Saargegend* 23, no. 4 (1975–76): 157.

30. Circular no. 65, 11 July 1892, in NEA.

31. Jürgen Kocka, "Bürgertum und bürgerliche Gesellschaft im 19. Jahrhundert. Europäische Entwicklungen und deutsche Eigenarten," in *Bürgertum im 19. Jahrhundert. Deutschland im europäischen Vergleich,* vol. 1, ed. Kocka (Munich: Deutscher Taschenbuch Verlag, 1988), 26–33; Kocka, "Bürgertum und Bürgerlichkeit als Problem der deutschen Geschichte vom späten 18. zum frühen 20. Jahrhundert," in *Bürger und Bürgerlichkeit im 19. Jahrhundert,* ed. Kocka (Göttingen: Vandenhoeck & Ruprecht, 1987), 21–63.

32. *Saarbrücker Zeitung* (hereafter *SZ*), 30 June 1891, no. 149, morning edition.

33. Speech in *SZ,* 22 September 1891, no. 221, evening edition.

34. See Stumm's comments during a bonus distribution ceremony in 1889, *SZ,* 10 July 1889, no. 158, evening edition; or a similar speech in *SZ,* 29 June 1893, no. 150, morning edition.

35. Speech in *Neue Saarbrücker Zeitung* (hereafter *NSZ*), 29 September 1896, no. 257, first edition, 2–3. For the "feudal" thesis, see Hellwig, *Carl Ferdinand,* 295; and Ascher, "Baron von Stumm," 279.

36. Tille, *Die Reden,* 12:588–89.

37. *SGB* summary of the paternalist regime, in Tille, *Die Reden,* 12:621–22.

38. "Arbeit als Besserungsmittel," *SZ,* 3 August 1889, no. 179, evening edition.

39. "Bestrafung der Arbeitsscheu," *Malstatt-Burbacher Zeitung* (hereafter *MBZ*), 12 October 1891, no. 238.

40. On the Enlightenment's positive revaluation of labor, see Werner Conze, "Arbeit," in *Geschichtliche Grundbegriffe. Historisches Lexikon zur politisch-sozialen Sprache in Deutschland,* vol. 1, ed. Otto Brunner, Werner Conze, and Reinhart Kosselleck (Stuttgart: E. Klett, 1972), 167–215; and William H. Sewell, *Work and Revolution in France: The Language of Labor from the Old Regime to 1848* (Cambridge: Cambridge University Press, 1980), 64ff.

41. Stumm's speech to workers in Neunkirchen, 21 June 1891, in Tille, *Die Reden,* 12:561; Hellwig, *Carl Ferdinand,* 298.

42. Stumm's speech during a factory ceremony in Neunkirchen, 21 June 1891, in Tille, *Die Reden,* 11:556; Boch's speech to his workforce during a fiftieth anniversary ceremony for the firm, June 1891, cited in *SZ,* 30 June 1891, no. 149, morning edition.

43. Stumm's speech to workers during a bonus distribution ceremony at the Neunkirch steelworks, in *SZ,* 10 July 1889, no. 158, evening edition.

44. The petition can be found in *SGB,* 14 October 1888, no. 42, 169–71. See also Fritz Hellwig, "Unternehmer und Unternehmungsform im saarländische Industriegebiet," *Jahrbuch für Nationalökonomie und Statistik* 158 (1943): 424.

45. Cited in Tille, *Die Reden,* 12:621.

46. See the factory circular dated 25 June 1887 in NEA.

47. Crew, *Town in the Ruhr,* 145–57; Lawrence Schofer, *The Formation of a Modern Labor Force: Upper Silesia, 1865–1914* (Berkeley and Los Angeles: University of California Press, 1975).

48. *SZ,* 1 June 1884, no. 127, first edition.

49. *SZ,* 30 June 1891, 149, morning edition.

50. *St. Johanner Zeitung* (hereafter *SJZ*), 6 July 1897, no. 155; Hellwig, *Carl Ferdinand.*

51. *SGB,* 27 April 1890, no. 17, 1–2.

52. On Stumm's rules, see his comments in *SGB,* 27 April 1890, no. 17, 1–2; Hellwig, *Carl Ferdinand,* 289ff.; Christoph Lang, "'Herren im Hause.' Die Unterhehmer," in *Industriekultur an der Saar. Leben und Arbeit in einer Industrieregion 1840–1914,* ed. Richard van Dülmen (Munich: Verlag C. H. Beck, 1990), 132–45. On the Burbach steelworks, see the company brochure entitled "Burbacher Hütte bei Saarbrücken. Wohlfahrtseinrichtungen 1897" in Stadtarchiv Saarbrücken (hereafter SAS), Bestand Bürgermeisterei Malstatt-Burbach, 379, 27–28. On Villery & Boch, see *SGB,* 21 June 1891, no. 25, 1.

53. "Wohlfahrtseinrichtungen für Fabrikarbeiter," 16.

54. *SZ,* 25 September 1890, no. 224, morning edition.

55. Junghahn, *Das Schlafhaus- und Einliegerwesen,* 1, 9.

56. Circular no. 26, 20 April 1885, in NEA. See also the *SGB* article cited in Tille, *Die Reden,* 12:621.

57. Max Vopelius, *Die Tafelglasindustrie im Saarthale* (Halle: n.p., 1895), 60–69, 71, 74.

58. The quotes come from the comments of BM Meyer of Malstatt-Burbach and Stumm, respectively, on company welfare schemes. See Meyer's draft letter from 1890 in SAS, Bestand Bürgermeisterei Malstatt-Burbach, 378; and Stumm's comments in circular no. 26, 20 April 1885, in NEA.

59. On the Burbach steelworks, see the company brochure entitled "Burbacher Hütte bei Saarbrücken. Wohlfahrtseinrichtungen 1897" in SAS, Bestand Bürgermeisterei Malstatt-Burbach, 379, 27–28. On Villery & Boch, see *SGB,* 21 June 1891, no. 25, 1.

60. For Stumm and Röchling, see Lang, "'Herren im Hause,'" 138. For Theodor Sehmer, see *SJZ,* 6 July 1897, no. 155. See the discussion in Hans-Jürgen Serwe, "Hausbau und Wohnen," in van Dülmen, *Industriekultur an der Saar,* 157–59, from which I have drawn the quotations in text.

61. Walter Lauer, "Geschichte der Maschinenfabrik Ehrhardt & Sehmer, Saarbrücken," *Zeitschrift für saarländische Heimatkunde* 2 (1952): 71–80.

62. BM (Malstatt-Burbach) to LR (Saarbrücken), 14 October 1884, and RP to LR (Saarbrücken), 5 December 1884, in LHAK, 442/6695, 55–58, 93–95.

63. Thomas Welskopp, "Betriebliche Sozialpolitik im 19. und frühen 20. Jahrhundert. Eine Diskussion neuerer Forschungen und Konzepte und eine Branchenanalyse der deutschen und amerikanischen Eisen- und Stahlindustrie von den 1870er bis zu den 1930er Jahren," *Archiv für Sozialgeschichte* 34 (1994): 334.

64. Johann Paul, *Alfred Krupp und die Arbeiterbewegung* (Düsseldorf: Schwann, 1987), 30–41, 96–121; Crew, *Town in the Ruhr,* 145–57.

65. Welskopp, "Betriebliche Sozialpolitik," 342. See also Wolfram Fischer, "Die Pionierrolle der betrieblichen Sozialpolitik im 19. und beginnenden 20. Jahrhundert," in *Betriebliche Sozialpolitik deutscher Unternehmen seit dem 19. Jahrhundert,* ed. Wilhelm Treue and Hans Pohl (Wiesbaden: Steiner, 1978), 46–50.

66. See Joël Michel, "Bergarbeiter-Kommunen und Patriarchalismus in Westeuropa vor 1914," in *Sozialgeschichte des Bergbaus im 19. und 20. Jahrhundert,* ed. Klaus Tenfelde (Munich: C. H. Beck, 1992), 58–63; Thomas Welskopp, *Arbeit und Macht im Hüttenwerk. Arbeits- und industrielle Beziehungen in der deutschen und*

amerikanischen Eisen- und Stahlindustrie von den 1860er bis zu den 1930er Jahren (Bonn: J. H. W. Dietz, 1994), 354ff.; Patrick Joyce, *Work, Society, and Politics: The Culture of the Factory in Later Victorian England* (New Brunswick, NJ: Rutgers University Press, 1980), especially 134–54; Sonya Rose, *Limited Livelihoods: Gender and Class in Nineteenth-Century England* (Berkeley and Los Angeles: University of California Press, 1992); Sanford Elwitt, *The Third Republic Defended: Bourgeois Reform in France, 1880–1914* (Baton Rouge and London: Louisiana State University Press, 1986), 114–69; and Donald Reid, *The Miners of Decazeville: A Genealogy of Deindustrialization* (Cambridge, MA: Harvard University Press, 1985).

67. E. P. Hennock, *The Origin of the Welfare State in England and Germany, 1850–1914* (Cambridge: Cambridge University Press, 2007); Hennock, "Social Policy under the Empire—Myths and Evidence," *German History* 16 (1998): 59–74.

68. Klaus-Michael Mallmann, "Die Anfänge der Sozialdemokratie im Saarrevier," *Zeitschrift für die Geschichte der Saargegend* 28 (1980): 128–32.

69. The announcement is published in Tille, *Die Reden,* 8:80ff. See the Saar Mining Office's notice to its workers entitled "Bergleute der Königlich Saarbrücker Gruben!" in LAS, 564/715, 96; and the report on Stumm's notification of his workers in *SJZ,* 11 July 1877, no. 158. See also Karl Alfred Gabel, *Kämpfe und Werden der Hüttenarbeiter-Organisationen an der Saar* (Saarbrücken: Saarbrücker Druckerei und Verlag, 1921), 70; Hellwig, *Carl Ferdinand,* 231.

70. The label comes from Gabel, *Kämpfe und Werden,* 71.

71. Horst Lademacher, "Wirtschaft, Arbeiterschaft, Arbeiterorganisationen in der Rheinprovinz 1877/78," *Archiv für Sozialgeschichte* 15 (1975): 140; Mallmann, "Die Anfänge," 134–35.

72. See the protocol of the 4 May 1885 meeting of the directorates of both organizations in *SGB,* 10 May 1885, no. 19, 73. On the formation of the Saar business organizations, see Fritz Hellwig, *Die Saarwirtschaft und ihre Organisationen. Seit der Errichtung der Industrie- und Handelskammer zu Saarbrücken 1863/64* (Saarbrücken: Buchgwerbehaus Aktiengesellschaft Saarbrücken, 1939), 55ff.

73. The resolution is reprinted in Tille, *Die Reden,* 8:81–82. See also Lademacher, "Wirtschaft, Arbeiterschaft, Arbeiterorganisationen," 118.

74. Lademacher, "Wirtschaft, Arbeiterschaft, Arbeiterorganisationen," 120–37.

75. Tille, *Die Reden,* 8:82.

76. See, e.g., Stumm's comments during the Reichstag debate over renewal in 1880 and the comments of Richard Vopelius in a Prussian Landtag debate in 1880 in Tille, *Die Reden,* 8:63–80, 92.

77. Monika Breger, *Die Haltung der industriellen Unternehmer zur staatlichen Sozialpolitik in den Jahren 1878–1891* (Frankfurt am Main: Haag & Herchen, 1982).

78. Hans-Peter Ullmann, "Industriellen Interessen und die Entstehung der deutschen Sozialversicherung 1880–1889," *Historische Zeitschrift* 229, no. 3 (1979): 589–91.

79. George Steinmetz, *Regulating the Social. The Welfare State and Local Politics in Imperial Germany* (Princeton, NJ: Princeton University Press), 126.

80. Ullmann, "Industriellen Interessen," 590.

81. Stumm first presented this motion as a delegate to the Reichstag of the

North German Confederation in 1867. See Hellwig, *Carl Ferdinand,* 211–19; Breger, *Die Haltung der industriellen Unternehmer,* 16; Ullmann, "Industrielle Interessen," 594–95.

82. See the description of the funds in Greg Eghigian, *Making Security Social: Disability, Insurance, and the Birth of the Social Entitlement State in Germany* (Ann Arbor: University of Michigan Press, 2000), 55.

83. See Stumm's comments from the debate over associations and class struggle in the Reichstag of the North German Confederation in 1867 in Tille, *Die Reden,* 8:25.

84. Steinmetz, *Regulating the Social,* 128.

85. Breger, *Die Haltung der industriellen Unternehmer,* 56–75; Hennock, *Origin of the Welfare State,* 158–60.

86. Ullmann, "Industrielle Interessen," 600–601. See also Hennock, *Origin of the Welfare State,* 96–100, which nevertheless downplays the interests of heavy industry.

87. See Breger, *Die Haltung der industriellen Unternehmer,* 136–39; and Ullmann, "Industrielle Interessen," 605–7. On the actions of Saar industrialists, see *SGB,* 25 December 1887, no. 52, 206, 208–9.

88. Jean Quataert, "A Source Analysis in German Women's History: Factory Inspectors' Reports and the Shaping of Working-Class Lives, 1878–1914," *Central European History* 16 (1983): 99–121.

89. Hennock, *Origin of the Welfare State,* 98.

90. Steinmetz, "The Myth of an Autonomous State: Industrialists, Junkers, and Social Policy in Imperial Germany," in *Society, Culture, and the State in Germany 1870–1930,* ed. Geoff Eley (Ann Arbor: University of Michigan Press, 1996), 299; Steinmetz, *Regulating the Social,* 33–35, 130.

91. Sabine Schmitt, *Der Arbeiterinnenschutz im deutschen Kaiserreich. Zur Konstruktion der schutzbedürftigen Arbeiterin* (Stuttgart and Weimar: J. B. Metzler, 1995), 86–91; Hans-Jörg Berlepsch, *"Neuer Kurs" im Kaiserreich? Die Arbeiterpolitik des Freiherrn von Berlepsch 1890 bis 1896* (Bonn: Verlag Neue Gesellschaft, 1987), 250–55.

92. This general argument follows Steinmetz's analyses in *Regulating the Social,* 80–107, and"Myth of an Autonomous State," 257–318.

93. For numerous examples of this, see Ullmann, "Industrielle Interessen," 594–95; and Breger, *Die Haltung der inudstriellen Unternehmer.*

94. For examples of Stumm's meetings with Bismarck, see Hellwig, *Carl Ferdinand,* 241, 145; for examples of his correspondence, see ibid., 249–88. On industrialists more generally, see Ullmann, "Industrielle Interessen," 595.

95. On the Volkswirtschaftsrat, which met for the last time in 1887, see Hellwig, *Carl Ferdinand,* 241–44; and Gerhard A. Ritter, "Politische Representation durch Berufsstände. Kozepte und Realität in Deutschland 1871–1933," in *Gestaltungskraft des Politischen. Festschrift für Eberhard Kolb,* ed. Wolfram Pyta and Ludwig Richter (Berlin: Duncker & Humblot, 1998), 265–68.

96. Ullmann, "Industrielle Interessen," 590.

97. Nicos Poulantzas, *State, Power, Socialism* (London: Verso, 2000), 136.

98. For the concerns of Saar industrialists, see their expert's report (*Gutachten*)

on the proposed old age and disability bill from 1887 in *SGB*, 25 December 1887, no. 52, 209. For industrialists generally, see Breger, *Die Haltung der industriellen Unternehmer*, 136–39; and Ullmann, "Industrielle Interessen," 605, 608.

CHAPTER 2

1. See Meyer's speech and the speech of Wilhelm Köhl, director of the Burbach steelworks rolling mill, in the report on the special meeting of the city council from 1 December 1891 in SAS, Bestand Bürgermeisterei Malstatt-Burbach, 1241.

2. Miriam Hansen, "Early Silent Cinema: Whose Public Sphere?" *New German Critique* 29 (1983): 156.

3. Michael Warner, "The Mass Public and the Mass Subject," in *Habermas and the Public Sphere*, ed. Craig Calhoun (Cambridge, MA: MIT Press, 1992), 377–401.

4. Jürgen Hannig, "Der beschleunigte Raum. Industrialisierung und Verkehr im Saarland," and Rolf Wittenbrock, "Industriedörfer und Verstädterung," in *Industriekultur an der Saar. Leben und Arbeit in einer Industrieregion 1840–1914*, ed. Richard van Dülmen (Munich: C. H. Beck, 1990), 27–41, 84–95; Stefan Leiner, *Migration und Urbanisierung. Binnenwanderungsbewegungen, räumlicher und sozialer Wandel in den Indusriestädten des Saar-Lor-Lux-Raumes 1856–1910* (Saabrücken: Saarbrücker Drückerei und Verlag, 1994); Albert Ruppersberg, *Geschichte der Gemeinde und Bürgermeisterei Dudweiler* (Saarbrücken: Gebr. Hofer A.-G., 1923); Ernst Klein, "Bergfiskus und Kirche an der Saar im 19. Jahrhundert," *Zeitschrift für die Geschichte der Saargegend* 23 (1975–76): 157–93.

5. *Protokoll-Buch in den Stadtverordneten Sitzungen* (hereafter *Protokoll-Buch*), meetings of 29 November 1900 and 18 November 1901, in SAS, Bestand Bürgermeisterei Malstatt-Burbach, 1246–47, unpaginated.

6. See the reports on the newly assigned factory directors of the Burbach steelworks in Malstatt-Burbach in 1901 and 1902, Johann Josef Ott and Edmund Weisdorff, in SAS, Bestand Bürgermeisterei Malstatt-Burbach, 1247, unpaginated.

7. William Harbutt Dawson, *Municipal Life and Government in Germany* (London: Longmans, 1914), especially 58–80; Heinrich Heffter, *Die Deutsche Selbstverwaltung im 19. Jahrhundert. Geschichte der Ideen und Institutionen* (Stuttgart: K. F. Koehler, 1950).

8. These figures are from the *Bericht über den Stand und die Verwaltung der Gemeinde-Angelegenheiten der Stadtgemeinde Malstatt-Burbach für das Rechnungsjahr 1876* (Malstatt-Burbach: Rußhütte Charlottenstiftung and Schmidtbornschen Stiftung, 1876).

9. Antje Fuchs, "Nach der Schicht. Zur Freizeitkultur von Hüttenarbeitern," in *Stumm in Neunkirchen. Unternehmererschaft und Arbeiterleben im 19. Jahrhundert. Bilder und Skizzen aus einer Industriegemeinde*, ed. Richard van Dülmen and Joachim Jacob (St. Ingbert: Röhrig Verlag, 1993), 131.

10. *Protokoll-Buch*, meeting of 27 November 1890, in SAS, Bestand Bürgermeisterei Malstatt-Burbach, 1240–41, unpaginated. On the Neunkrichen case, see

the undated report compiled by the major of Neunkirchen and the report on the 22 April 1897 council meeting in "Versammlungen und Geschäfte des Bürgermeistereirats und der Gemeinderäte," in Stadtarchiv Neunkirchen (hereafter SAN), AI, 9, 29–31.

11. *SZ,* 21 February 1885, no. 44.

12. See *Neunkircher Adreßbuch, enthaltend die Einwohner der Orte Neunkirchen, Niederneunkirchen, Schlaweries, Sinnerthal, Heinitz-Dechen, Kohlhof, Spiesen, Elversberg und Wellesweiler* (Neunkirchen: Buchdruckerei von C. A. Ohle, 1888), 15; compare *Adreßbuch von Neunkirchen 1910* (Neunkirchen: C. A. Ohle, 1910), 4.

13. See the list of council members from 1903 until 1909 in LAS, Depositum Sulzbach, Fach 22, no. 8, unpaginated.

14. *Saarabia vor Gericht. Bericht über den Prozeß Hilger gegen Krämer unter Benutzung stenographischer Aufzeichnungen* (Berlin: Verlag der Buchhandlung Vorwärts. 1904), 42.

15. *SZ,* 13 November 1893, no. 267, evening edition; *SZ,* 18 December 1893, no. 296, evening edition.

16. See the list in SAS, Bestand Bürgermeisterei Malstatt-Burbach, 1244, unpaginated.

17. *SZ,* 5 October 1897, no. 272, first edition; *Protokoll-Buch,* meeting of 14 January 1889, in SAS, Bestand Bürgermeisterei Malstatt-Burbach, file 1240; *Die Burbacher Hütte 1856–1906. Denkschrift zur Feier des fünfzigjährigen Bestehens der Hütte am 22. Juni 1906* (Saarlouis: Franz Stein Nachf. Hausen & Co., 1906), 102; *Protokoll-Buch,* meeting of 14 January 1889, in SAS, Bestand Bürgermeisterei Malstatt-Burbach, 1240.

18. Fuchs, "Nach der Schicht," 134–36.

19. *Protokoll-Buch,* meetings of 3 April 1906 and 2 May 1907, in SAS, Bestand Bürgermeisterei Malstatt-Burbach, 1249, unpaginated.

20. See the notice posted on the factory gate and published in *MBZ,* 31 December 1896, no. 306; and the company's official notice in SAS, Bestand Bürgermeisterei Malstatt-Burbach, 215, unpaginated.

21. BM Meyer (Malstatt-Burbach) to police commissioner of Malstatt-Burbach, 15 October 1887, in SAS, Bestand Bürgermeisterei Malstatt-Burbach, 215, unpaginated; BM Meyer to the Burbach steelworks management, 7 November 1887, in ibid.; factory director Hans Seebohm to BM Meyer, 10 November 1887, in ibid.

22. *MBZ,* 2 January 1897, no. 1.

23. "Feier von Kirmesfesten 1885–1902" and *Protokoll-Buch,* meeting of 20 November 1884, in SAS, Bestand Bürgermeisterei Malstatt-Burbach, 1138, unpaginated.

24. BM (Malstatt-Burbach) to Burbach steelworks, 27 May 1889, in "Feier von Kirmesfesten 1885–1902," in SAS, Bestand Bürgermeisterei Malstatt-Burbach, 1138, unpaginated; Hans Seebohm and Pequin of Burbach steelworks to BM, 3 June 1889, in ibid.; BM to Burbach steelworks, 27 June 1889 (which discusses the council's decision), in ibid.

25. *SZ,* 11 June 1897, no. 156.

26. *Protokoll-Buch,* meeting of 12 May 1898, in SAS, Bestand Bürgermeisterei

Malstatt-Burbach, 1245; *St. Johanner-Saarbrücker Volkszeitung* (hereafter *SJSVZ*), 16 June 1897, no. 136, in "Feier von Kirmesfesten 1885–1902," in SAS, Bestand Bürgermeisterei Malstatt-Burbach, 1138, unpaginated.

27. See the resolution reprinted in Alexander Tille, ed., *Die Reden des Freiherrn Carl Ferdinand von Stumm-Halberg,* vol. 12 (Berlin: Otto Elsner, 1915), 594–95.

28. See Saul, "Der Staat und die 'Mächte des Umsturzes.' Ein Beitrag zu den Methoden antisozialistischer Repression und Agitation vom Scheitern des Sozialistengesetzes bis zur Jahrhundertwende," *Archiv für Sozialgeschichte* 12 (1972): 293–350.

29. See, e.g., the report in LR (Ottweiler) to BM (Neunkirchen), 13 January 1891, in "Baupolizei—Versammlungsräume 1889–1896," in SAN, AI, 143, 79–80; and the various exchanges with a widow Kohler from Spiesen and a publican Wolter in ibid., 97, 9–10, 68–70, 61–62.

30. The last is mentioned in *Saar- und Blieszeitung,* 12 March 1898, no. 60, second edition, in SAN, AI, 9.

31. *Protokoll-Buch,* meeting of 8 August 1896, in SAS, Bestand Bürgermeisterei Malstatt-Burbach, 1244, unpaginated.

32. *General-Anzeiger* (hereafter *GA*), 21 June 1890, no. 142, in SAS, Bestand Bürgermeisterei Malstatt-Burbach, 378, unpaginated.

33. For the mayor's comments, see his report in SAS, Bestand Bürgermeisterei Malstatt-Burbach, 378, unpaginated; *MBZ,* 21 March 1891, no. 68; the letter from Dr. Brüggemeier to BM Meyer, 6 July 1890, in ibid.; *Protokoll-Buch,* meeting of 19 June 1890 (the council meeting in which the idea was first broached), in SAS, Bestand Bürgermeisterei Malstatt-Burbach, 1240.

34. Fritz Hellwig, *Carl Ferdinand von Stumm-Halberg 1836–1901* (Heidelberg and Saarbrücken: Westmark-Verlag, 1936), 323–26; Christoph Lang, "'Herren im Hause.' Die Unternehmer," in *Industriekultur an der Saar. Leben und Arbeit in einer Industrieregion 1840–1914,* ed. Richard van Dülmen (Munich: Verlag C. H. Beck, 1990), 145.

35. *Protokoll-Buch,* meetings of 6 October 1892 and 12 January 1893, in SAS, Bestand Bürgermeisterei Malstatt-Burbach, 1241. Officials of the Burbach steelworks managed to ensure that they were well represented on the committee charged with drawing up the statutes of the new *Gewerbegericht* (*Protokoll-Buch,* meetings of 12 January 1893 and 9 February 1893, in ibid.).

36. See the reports in LR (Ottweiler) to RP, 24 January 1893, in LHAK, 442/4256, 581–84; and LR (Ottweiler) to RP, 25 February 1893, in LHAK, 442/4402, 77.

37. *SZ,* 13 November 1893, no. 267, evening edition; *SZ,* 18 December 1893, no. 296, evening edition.

38. David Blackbourn, "The Discreet Charm of the Bourgeoisie: Reappraising German History in the Nineteenth Century," in Blackbourn and Geoff Eley, *The Peculiarities of German History* (Oxford: Oxford University Press, 1984), 192ff.; Eley, "Nations, Publics, and Political Cultures: Placing Habermas in the Nineteenth Century," in *Habermas and the Public Sphere,* ed. Craig Calhoun (Cambridge, MA: MIT Press, 1992), 296–98.

39. *Statuten der Saarbrücker-Casino-Gesellschaft* (Saarbrücken: Georg Pfeiffer,

1860), 7; *Statuten der Saarbrücker Casino-Gesellschaft* (Saarbrücken: Gebrüder Hofer, 1887), 5; *Saarbrücker Kasino-Chronik 1796–1896* (Saarbrücken: Gebrüder Hofer, 1896), 49–51; *Kasino-Gesellschaft Saarbrücken. Mitglieder-Verzeichnis, 1. April 1905* (Saarbrücken: Gebrüder Hofer, 1905).

40. Fr. Pelzer to BM Meyer, 10 December 1875, in SAS, Bestand Bürgermeisterei Malstatt-Burbach, 311, 1.

41. Fr. Pelzer to BM Meyer, 10 December 1875, in SAS, Bestand Bürgermeisterei Malstatt-Burbach, 311, 1; the statutes and membership list of the association in SAS, Bestand Alt-Saarbrücken, 2072, 2–5.

42. *Statuten des Casino-Vereins,* 4 December 1853, in Stadtarchiv Dudweiler (hereafter SAD), 419, unpaginated.

43. *Statuten der Neunkircher Casino-Gesellschaft, vom 1. März 1873* (Neunkirchen: Buchdruckerei von C. A. Ohle, 1873); *Neunkircher Adreßbuch* (Neunkirchen: Buchdruckerei von C. A. Ohle, 1888), 215; Fuchs, "Nach der Schicht, 94.

44. Blackbourn and Eley, *Peculiarities of German History,* 225.

45. For the kaiser's birthday celebration, see, e.g., *SZ,* 20 March 1883, no. 66; for the *Jubiläumsfeier* for Raabe, see *SZ,* 5 October 1897, no. 272, first edition; for the housing cooperative, see the report in *MBZ,* 21 March 1891, no. 68.

46. See the 1872 statutes and membership lists of the *Bürger-Verein,* originally formed in 1862, in SAS, Bestand Alt-Saarbrücken, 1733, 16.

47. See the lists in *Adreßbuch für die Städte St. Johann a. d. Saar und Saarbrücken,* comp. Stadt-Sekretär Bier and Stadt-Sekretär König (St. Johann and Saarbrücken: privately printed, 1888), 343, 346.

48. *SZ,* 23 March 1884, no. 71; *SZ,* 27 April 1884, no. 99.

49. *SZ,* 26 March 1885, no. 72; *SJZ,* 19 January 1898, no. 15.

50. See the lecture schedule for the winter 1912–13 season in LAS, Depositum Sulzbach, Fach 68, no. 4, unpaginated.

51. For the statutes and membership of the music association, see SAS, Bestand Bürgermeisterei Malstatt-Burbach, 311, unpaginated; and SAS, Bestand Alt-Saarbrücken, 1817, 15, 16, 19.

52. Thomas Rohkrämer, *Der Militärismus der "kleinen Leute." Die Kriegervereine im Deutschen Kaiserreich* (Munich: R. Oldenbourg, 1990); Klaus Saul, "Der 'Deutsche Kriegerbund.' Zur innenpolitischen Funktion eines 'nationalen' Verbandes im kaiserlichen Deutschland," *Militärgeschichtliche Mitteilungen* 1 (1971): 95–159; H. Henning, "Kriegervereine in den preussischen Westprovinzen. Ein Beitrag zur preussischen Innenpolitik zwischen 1860 und 1914," *Rheinische Vierteljahresblätter* 32 (1968): 430–75.

53. "Geschäfts-Bericht des Bezirks-Vorstandes für die Zeit vom 1. January 1888 bis Ende Dezember 1891. Deutscher Kriegerbund. Saar-, Blie- und Nahe-Bezirk," in LHAK, 442/11162, unpaginated.

54. On the first two examples, from 1891 and 1894, see SAD, 508, unpaginated. On the second two examples, see, respectively, SAS, Bestand Alt-Saarbrücken, 1878, unpaginated; and LAS, Depositum Sulzbach, Fach 68, no. 4, vol. 1, unpaginated.

55. *SJZ,* 19 July 1893, no. 167.

56. LR (Saarbrücken) to the BMs of the Kreis, 30 July 1904, in LAS, Depositum Sulzbach, Fach 68, no. 5, unpaginated; "concluding remarks" in "Geschäfts-Bericht des Bezirks-Vorstandes für die Zeit vom 1. January 1888 bis Ende Dezember 1891. Deutscher Kriegerbund. Saar-, Blie- und Nahe-Bezirk," in LHAK, 442/11162, unpaginated.

57. *SZ,* 2 March 1908, no. 52, second edition.

58. "Frauenverein zu Saarbrücken. Bericht 1860," in LHAK, 403/7364, 27; *Der Bergmannsfreund* (hereafter *BF*) 30 March 1905, no. 37, 295–96; *SZ,* 23 February 1891, no. 45, evening edition.

59. See the list in *Adreßbuch für die Städte St. Johann, Saarbrücken, Malstatt-Burbach und Umgebung 1905* (Saarbrücken: Druck und Verlag der Saardruckerei, 1905), 451.

60. *SZ,* 24 December 1890, no. 301, evening edition.

61. *SZ,* 2 March 1908, no. 52, second edition, in SAS, Bestand Bürgermeisterei Malstatt-Burbach, 312.

62. See the reports and statutes in *Evangelisches Wochenblatt* (hereafter *EW*), 27 January 1889, no. 4, 31; *EW,* 10 February 1889, no. 6, 44–45; *SZ,* 9 July 1902, no. 185, in SAS, Bestand Alt-Saarbrücken, 1755, 28; and *NSZ,* 16 July 1901, no. 177, 2–3. For the Protestant League's sociology, see the first membership list from March 1889 in SAS, Bestand Alt-Saarbrücken, 1755, 30–31. See, more generally, Helmut Walser Smith, *German Nationalism and Religious Conflict: Culture, Ideology, Politics, 1870–1914* (Princeton, NJ: Princeton University Press, 1995).

63. Ludwig Lindsmayer, "Geselligkeit und Selbstbestimmung. Die Vereinskultur," in van Dülmen, *Industriekultur an der Saar,* 226.

64. Ibid.; Klaus-Michael Mallmann, *Die Anfänge der Bergarbeiterbewegung an der Saar (1848–1904)* (Saarbrücken: Minerva-Verlag Thinnes & Nolte, 1981), 51–55.

65. Lists of Catholic organizations can be found for St. Johann and Saarbrücken in *Adreßbuch für die Städte St. Johann a. d. Saar und Saarbrücken sowie Umgegend* (Saarbrücken and St. Johann: Im Selbstverlage der Verfasser, 1888), 324–26, and for Neunkirchen in *Neunkircher Adreßbuch* (Neunkirchen: Buchdruckerei von C. A. Ohle, 1888), 215.

66. Josef Bellot, *Hundert Jahre politisches Leben an der Saar unter preussischer Herrschaft (1815–1918)* (Bonn: Ludwig Röhrscheid Verlag, 1954).

67. *SZ,* 5 June 1893, no. 129, morning edition.

68. See the membership list of the National Liberal electoral association for the district of Saarbrücken from 1885 in SAS, Bestand Alt-Saarbrücken, 1728, unpaginated.

69. See the National Liberal *Wahlaufruf* in *SZ,* 28 October 1884, no. 253; the National Liberal *Wahlaufruf* in *SZ,* 20 February 1887, no. 43, first edition; and the list of executive committee members for the district in Secretary Kriene to LR (Saarbrücken), 9 February 1885, in SAS, Bestand Alt-Saarbrücken, 1728, unpaginated.

70. See the 1887 electoral handbill entitled "An die Wähler zum deutschen Reichstage im Wahlkreis Ottweiler—St. Wendel—Meisenheim" in NEA, 390, unpaginated.

71. I have matched party affiliations from campaign manifestos with the list of councillors in "Protocoll-Buch in den Stadtverordneten Sitzungen 1888–1891," meeting of 19 June 1890, in SAS, Bestand Bürgermeisterei Malstatt-Burbach, 1240, unpaginated.

72. In this case, I have compared the memberships in the National Liberal and Free Conservative electoral committee of 1887 listed in NEA, 390 (unpaginated), with the municipal council of Neunkirchen in 1888 listed in *Neunkircher Adreßbuch* (Neunkirchen: Buchdruckerei von C. A. Ohle, 1888), 15.

73. Nipperdey, *Die Organisation der deutschen Parteien vor 1918* (Düsseldorf: Droste Verlag, 1961), 42–85, 86–108. See also Geoff Eley, *Reshaping the German Right: Radical Nationalism and Political Change after Bismarck* (New Haven: Yale University Press, 1980), 19–40.

74. For just one example of these private meetings, see the report on the trustee meeting held by Free Conservatives and National Liberals in Stumm's casino in Neunkirchen on 27 March 1898 in *SZ,* 29 March 1898, no. 86, first edition. The meeting, led by Dr. Fuller, was an attempt to convince Stumm to run in the Reichstag election.

75. "Wahlaufruf der nationalliberalen Partei," *NSZ,* 3 May 1898, no. 118, 1; "Zur Stellung der Nationalliberalen im Gebiete," *SZ,* 10 February 1885, no. 34; "Wahlaufruf der Nationalliberalen Partei," *NSZ,* 5 May 1903, no. 105, 1, and 6 May 1903, no. 106, 1.

76. *SZ,* 10 February 1885, no. 34; *SZ,* 19 April 1885, no. 91, first edition.

77. On "apoliticism," see Rudy Koshar, *Social Life, Local Politics, and Nazism: Marburg, 1880–1935* (Chapel Hill: University of North Carolina Press, 1986), 6, 73, 75, 276–77.

78. *SJZ,* 3 October 1893, no. 232, morning edition.

79. "Zur Stellung der Nationalliberalen im Saargebiete," *SZ,* 10 February 1885, no. 34.

80. Bericht der Wahlprüfungs-Kommission über die Wahl des Abgeordneten Boltz im 5. Wahlkreise des Regierungsbezirks Trier (Saarbrücken), excerpt from Stenographische Berichte über die Verhandlungen des Reichstages, 11. Legislaturperiode, 1. Session, 1903/4, in LAS, Landratsamt Saarbrücken, Wa17, unpaginated.

81. This factory gate notice was dated 14 June 1903 and signed by steelworks director Edmund Weisdorff. See *SZ,* 15 June 1903, no. 161.

82. Cited in *SZ,* 3 March 1887, no. 52.

83. See the "election appeal" (*Wahlaufruf*) of the electoral committee of Ottweiler-St. Wendel-Meisenheim printed in *SZ,* 15 March 1889, no. 63, evening edition.

84. "An die Arbeiter!" in NEA, 390 (emphasis in original).

85. Stumm to *Beamten,* dated 17 June 1887, in NEA, 390.

86. "Disciplinar-Reglement für die Meister der Firma Gebrüder Stumm zu Neunkircher Eisenwerk," 1 May 1902, in NEA.

87. This official policy was openly admitted by Mining Office chairman Ewald Hilger in 1904; see *Saarabia vor Gericht,* 88.

88. The list of chairmen and vice chairmen for the twelve polling stations in Malstatt-Burbach is printed in *MBZ,* 4 June 1903, no. 128.

89. *Saarabia vor Gericht,* 23.

90. For just two examples of active initiative from local officials, see the reports on the exchanges between the *Landrat* of Saarbrücken and Mining Office officials in LR (Saarbrücken) to RP, 27 November 1893, in LHAK, 442/3793, 469–78; and the collusion between the *Landrat* of Saarlouis, the mayor of Wallerfangen, and René von Boch, owner of the Villeroy & Boch ceramics works in Wallerfangen, in LR (Saarlouis) to RP, 19 March 1890, LHAK, 442/6694.

91. *Saarabia vor Gericht,* 79.

92. Ibid., 20–21, 27; "Auszug aus der Wahlanfechtungsschrift des Redakteurs Hoffmanns in St. Johann vom 1 Dezember 1898," in LHAK, 442/6589, 98. For a study of National Liberal electoral influence and manipulation at the level of regional politics in Prussia, see Thomas Kühne, "Die Liberalen bei den preussischen Landtagswahlen im Kaiserreich. Wahlmanipulation, Lokalismus und Wahlkompromisse," in *Liberalismus und Region. Zur Geschichte des deutschen Liberalismus im 19. Jahrhundert,* ed. Deiter Langewiesche and Lothar Gall (Munich: Oldenbourg, 1995), 277–305. On the activities of liberal and conservative industrialists nationally, see Margaret Lavinia Anderson, *Practicing Democracy: Elections and Political Culture in Imperial Germany* (Princeton, NJ: Princeton University Press, 2000), 199–237.

93. *SZ,* 29 June 1893, no. 150, morning edition.

94. RP to OP, 27 May 1893, in LHAK, 442/6656, unpaginated.

95. See Stumm's comments during a bonus distribution to the employees of his steelworks in Neunkirchen in *SZ,* 29 June 1893, no. 150, morning edition.

96. Ulrich Fohrmann, *Trierer Kulturkampfpublizistik im Bismarckreich. Leben und Werk des Preßkaplans Georg Friedrich Dasbach* (Trier: Paulinus-Verlag, 1977): Mallmann, *Die Anfänge,* 90–92; Klaus-Michael Mallmann, "Die neue Attraktivität des Himmels. Kirche, Religion und industrielle Modernisierung," in van Dülmen, *Industriekultur an der Saar,* 255; and Mallmann, "'Aus des Tages Last machen sie ein Kreuz des Herrn . . . ?' Bergarbeiter, Religion und sozialer Protest im Saarrevier des 19. Jahrhunderts," in *Volksreligiosität in der modernen Sozialgeschichte,* ed. Wolfgang Schieder (Göttingen: Vandenhoeck & Ruprecht, 1986), 168–69.

97. Bellot, *Hundert Jahre,* 131, 149.

98. Mallmann, *Die Anfänge,* 92.

99. Bellot, *Hundert Jahre,* 147, 155

100. Ibid., 147, 175.

101. The list also included a number of artisans, publicans, and farmers. See the appeal and list in LHAK, 403/8460, unpaginated.

102. Bellot, *Hundert Jahre,* 161–62.

103. Suval, *Electoral Politics in Wilhelmine Germany* (Chapel Hill: University of North Carolina Press, 1985), 37–54; Anderson, *Practicing Democracy.*

CHAPTER 3

1. Oscar Negt and Alexander Kluge, *Public Sphere and Experience: Toward an Analysis of the Bourgeois and Proletarian Public Sphere* (Minneapolis: University of Minnesota Press, 1993), 3–8, 57–63, especially 63; Nancy Fraser, "Rethink-

ing the Public Sphere: A Contribution to the Critique of Actually Existing Democracy," in *Justice Interruptus: Critical Reflections on the "Postsocialist" Condition* (New York and London: Routledge, 1997), 81.

2. Fritz Hellwig, *Carl Ferdinand von Stumm-Halberg 1836–1901* (Heidelberg and Saarbrücken: Westmark-Verlag, 1936), 506ff.; John C. G. Röhl, *Germany without Bismarck: The Crisis of Government in the Second Reich, 1890–1900* (London: B. T. Batsford, 1967), 120ff.

3. This point about Stumm's "catalytic effect" is made in Kevin Repp, *Reformers, Critics, and the Paths of German Modernity: Anti-Politics and the Search for Alternatives, 1890–1914* (Cambridge, MA: Harvard University Press, 2000), 259.

4. Karl Erich Born, *Staat und Sozialpolitik seit Bismarcks Sturz* (Wiesbaden: F. Steiner, 1957), 28–30; Hans-Jörg Berlepsch, *"Neuer Kurs" im Kaiserreich? Die Arbeiterpolitik des Freiherrn von Berlepsch 1890 bis 1896* (Bonn: Verlag Neue Gesellschaft, 1987), 17–19. On Bismarck, see Wilhelm Mommsen, *Bismarcks Sturz und die Parteien* (Stuttgart, Berlin, and Leipzig: Deutsche Verlags-Anstalt, 1924).

5. The complete text of the February Decrees is reprinted in Berlepsch, *"Neuer Kurs,"* 28–29.

6. Born, *Staat und Sozialpolitik,* 98–105; Berlepsch, *"Neuer Kurs";* George Steinmetz, *Regulating the Social: The Welfare State and Local Politics in Imperial Germany* (Princeton, NJ: Princeton University Press, 1993), 108–45; E. P. Hennock, *The Origin of the Welfare State in England and Germany: Social Policies Compared* (Cambridge: Cambridge University Press, 2007): 121–26.

7. Monika Breger, *Die Haltung der industriellen Unternehmer zur staatlichen Sozialpolitik in den Jahren 1878–1891* (Frankfurt am Main: Haag & Herchen, 1982), 159ff.; Berlepsch, *"Neuer Kurs,"* 256–61; Sabine Schmitt, *Der Arbeiterinnenschutz im deutschen Kaiserreich* (Stuttgart: J. B. Metzler, 1995), 94, 98–100.

8. Berlepsch, *"Neuer Kurs,"* 294–300.

9. Ibid., 309.

10. Klaus-Michael Mallmann, *Die Anfänge der Bergarbeiterbewegung an der Saar (1848–1904)* (Saarbrücken: Minerva-Verlag Thinnes & Nolte, 1981), 189–96.

11. *SZ,* 16 July 1889, no. 163, morning edition.

12. Horst Steffens, *Autorität und Revolte. Alltagsleben und Streikverhalten der Bergarbeiter an der Saar im 19. Jahrhundert* (Weingarten: Drumlin, 1987), 66.

13. For the figures, see Mallmann, *Die Anfänge,* 121–23; and Peter Kiefer, *25 Jahre Gewerkverein christlicher Bergarbeiter im Saarrevier* (Saarbrücken: Verlag des Gewerkvereins christlicher Bergarbeiter, 1929), 16. The percentages are taken from Steffens, *Autorität,* 47.

14. Kiefer, *25 Jahre Gewerkverein,* 14–15; Mallmann, *Die Anfänge,* 106–7.

15. "Abänderung für die königlichen Steinkohlengruben bei Saarbrücken vom 6. August 1877," in LAS, 564, 104–5.

16. Steffens, *Autorität,* 53–54.

17. Ibid., 66.

18. Peter Kiefer, *Die Organisationsbestrebungen der Saarbergleute, ihre Ursachen und Wirkungen auf dem Bereich des Saarbrücker Bergbaues und ihre Berechtigung* (Sulzbach-Saar: Peter Kiefer, 1912), 221.

19. Oberwachtmeister Stephan to LR (Ottweiler), 28 July 1890, in LHAK, 442/4304, 385; Roll to LR (Saarbrücken), 2 June 1890, in SAS, Bestand Bürgermeiterei Malstatt-Burbach, no. 727.

20. LR (Saarbrücken) to RP, 28 July 1890, in LAS, Landratsamt Saarbrücken, S/6; LR (Saarbrücken) to RP, 6 June 1890, in LHAK, 442/4420, 379–81; the two lists in SAS, Bestand Bürgermeisterei Malstatt-Burbach, no. 727.

21. Oberwachtmeister Stephan to LR (Ottweiler), 28 July 1890, in LHAK, 442/4304, 384–85.

22. *SGB,* 2 June 1890, no. 22, 1.

23. For a copy of this notice, see SAS, Bestand Bürgermeisterei Malstatt-Burbach, no. 727. Announcements for both firms were reprinted in *SGB,* 8 June 1890, no. 23, 1–2.

24. Seebohm to BM (Malstatt-Burbach), 14 June 1890, in SAS, Bestand Bürgermeisterei Malstatt-Burbach, no. 727.

25. *SJZ,* 1 July 1890, no. 151; Oberwachtmeister Stephan to LR (Ottweiler), 28 July 1890, in LHAK, 442/4304, 384–85; *SJZ,* 31 July 1890, no. 177; *SZ,* 30 October 1890, no. 254, morning edition; *SZ,* 31 October 1890, no. 255, evening edition.

26. Roll to LR (Saarbrücken), 2 June 1890, in SAS, Bestand Bürgermeisterei Malstatt-Burbach, no. 727; LR (Saarbrücken) to BM (Malstatt-Burbach), 14 June 1890, in SAS, Bestand Bürgermeisterei Malstatt-Burbach, no. 727.

27. Roll to police (St. Johann), 9 October 1891, in LHAK442/4274; Mallmann, *Die Anfänge,* 153.

28. Mallmann, *Die Anfänge,* 242–49.

29. Warken managed to win 6,823 votes (against the National Liberal Pfaehler's 10,214 and the Center candidate Schaede's 3,122), and Bachmann won 2,591 (against Stumm's 13,837 and Dasbach's 9,768) See Josef Bellot, *Hundert Jahre politisches Leben an der Saar unter preussischer Herrschaft (1815–1918)* (Bonn: Ludwig Röhrscheid Verlag, 1954), 181–85.

30. Klaus-Michael Mallmann, "'Saufkasinos' und Konsumvereine. Zur Genossenschaftsbewegung der Saarbergleute 1890–1914," *Der Anschnitt. Zeitchrift für Kunst und Kultur im Bergbau* 1 (1980): 200–206. For the seminal study of similar developments in the Ruhr, see Lutz Niethammer and Franz-Josef Brüggemeier, "Schlafgänger, Schapskasinos und schwerindustrielle Kolonie," in *Fabrik—Familie—Feierabend. Beiträge zur Sozialgeschichte des Alltags im Industriezeitalter,* ed. Jürgen Reulecke and Wolfhard Weber (Wuppertal: Peter Hammer Verlag, 1978), 158ff.

31. Cited in Steffens, *Autorität,* 72, 73.

32. On the prosecution of Warken, see Mallmann, *Die Anfänge,* 161–66. For ARV chairman Roll's similar legal troubles, see *SJZ,* 31 July 1890, no. 177; *SZ,* 30 October 1890, no. 254; and *SZ* 31 October 1890, no. 255.

33. Steffens, *Autorität,* 55.

34. Mallmann, *Die Anfänge,* 233–36.

35. Steffens, *Autorität,* 312.

36. Ibid., 46 and especially 67–84.

37. Bernd Warnecken, "'Massentritt.' Zur Körpersprache von Demonstranten in Kaiserreich," *Transformationen der Arbeiterkultur. Beiträge der 3. Arbeitstagung*

der Kommission "Arbeiterkultur" in der Deutschen Gesellschaft für Volkskunde in Marburg vom 3. bis 6. Juni 1985, ed. Peter Assion (Marburg: Jonas, 1996), 64–79.

38. On the 1892–93 strike, see Steffens, *Autorität,* 85–104; and Mallmann, *Die Anfänge,* 288–96.

39. Klaus-Michael Mallmann, "'Haltet fest wie der Baum die Äst.' Zur Rolle der Frauen in der Bergarbeiterbewegung 1892/93," *Saarheimat. Zeitschrift für Kultur, Landschaft, Volkstum* 24 (1980): 89.

40. *SZ,* 6 January 1893, no. 5, morning edition; *SZ,* 9 January 1893, no. 7, morning edition.

41. Cited in Steffens, *Autorität,* 99; and Horst Steffens, "Einer für alle, alle für einen? Bergarbeiterfamilien in der 2. Hälfte des 19. Jahrhunderts," in *Haushalt und Verbrauch in historischer Perspektive. Zum Wandel des privaten Verbrauchs in Deutschland im 19. und 20. Jahrhundert,* ed. Toni Pierenkemper (St. Katharinen: Scripta Mercaturae, 1987), 223–34.

42. Mallmann, *Die Anfänge,* 302.

43. Ibid., 288ff.

44. Cited in Michael Sander, "Katholische Arbeitervereine Berliner Richtung," *Archiv für mittelrheinische Kirchengeschichte* 37 (1985): 118.

45. Wilfried Loth, *Katholiken im Kaiserreich. Der politische Katholizismus in der Krise des wilhelminischen Deutschlands* (Düsseldorf: Droste Verlag, 1984), 82–83: Hostwalter Heitzer, *Der Volksverein für das katholische Deutschland im Kaiserreich 1890–1918* (Mainz: Matthias Brünewald, 1979), 15ff.

46. See the statutes of the Volksverein in LHAK, 442/6383, 279; and the 1893 election handbill entitled "Katholische Bergleute des Saarreviers!" in LHAK, 442/4402.

47. BM (St. Johann) to LR (Saarbrücken), 22 August 1892, in LAS, Landratsamt Saarbrücken, S/7, unpaginated; *SJZ,* 22 August 1892, no. 195.

48. Heitzer, *Der Volksverein,* 23.

49. BM (St. Johann) to LR (Saarbrücken), 22 August, 1892, in LAS, Landratsamt Saarbrücken, S/7, unpaginated; *SJZ,* 22 August 1892, no. 195.

50. LR (Saarbrücken) to RP, 27 November 1892, in LAS, Landratsamt Saarbrücken, S/7, and LHAK, 442/4250, 335; LR (Saarbrücken) to RP, 15 March 1894, in LHAK, 442/6383, 269–71.

51. BM (Heusweiler) to LR (Saarbrücken), 26 August 1892, and BM (Püttlingen) to LR (Saarbrücken), 25 August 1892, in LAS, Landratsamt Saarbrücken, S/7, unpaginated. This perspective draws on Michel Foucault's seminal discussions of "pastoral" power: see Foucault, "Politics and Reason," in *Michel Foucault: Politics, Philosophy, Culture; Interviews and Other Writings,* ed. with an introduction by Laurence D. Kritzman (New York and London: Routledge, 1988), 57–85.

52. See the conference report in SAS, Bestand Alt-Saarbrücken, 1423, 6–7.

53. Cited in Mallmann, *Die Anfänge,* 222.

54. "Katholische Bergleute des Saar-Reviers!" in LHAK, 442/4420, 411.

55. *SJZ,* 21 October 1896, no. 248; LR (Saarbrücken) to RP, 28 February 1895, and RP to OP, 3 March 1895, in LHAK, 442/4371, 356–62, 370–72.

56. The figures are taken from *SJZ,* 23 September 1895, no. 222; and Michael Sander, "Katholische Geistilichkeit und Arbeiterorganisation 2 Gutachten aus

dem Saarrevier und die Vorgeschichte des 'Fuldaer Pastorale' von 1900," in *Soziale Frage und Kirche im Saarrevier. Beiträge zu Sozialpolitik und Katholizismus im späten 19. und frühen 20. Jahrhundert* (Saarbrücken: Saarbrücker Druckerei und Verlag, 1984), 286.

57. *SJZ,* 16 February 1895, no. 40; Sander, "Katholische Geistilichkeit und Arbeiterorganisation," 286.

58. The quotes are taken from section 2 of the league's statutes, in LHAK, 403/6840, 677–92.

59. See section 3 of the league's statutes, in LHAK, 403/6840, 677–92; and the report on the events at the annual league festival in 1895 in *SJZ,* 23 September 1895, no. 222.

60. "An das katholische deutsche Volk!" in LHAK, 442/6383, 283–85.

61. Ibid.

62. The quotes are from Volksverein director Dr. Hohn, from Mönchengladbach, and his lecture on aims and goals of the Volksverein during a meeting in St. Johann on 28 Feb. 1904, in *SJSVZ,* 1 March 1904, no. 49, first edition.

63. For works that stress this oppositional dimension of Catholic organizing, see Claudia Hiepel, *Arbeiterkatholizismus an der Ruhr. August Brust und der Gewerkverein christlicher Bergarbeiter* (Stuttgart: W. Kohlhammer, 1999); and Margaret Livinia Anderson, *Practicing Democracy: Elections and Political Culture in Imperial Germany* (Princeton, NJ: Princeton University Press, 2000).

64. See the reports in BM (Sulzbach) to LR (Saarbrücken), 30 August 1892, and LR (Saarbrücken) to RP, 27 November 1892, in LAS, Landratsamt Saarbrücken, S/7, unpaginated. The personal connections and explicit party-political affiliations of Volksverein leaders are discussed in Heitzer, *Der Volksverein,* 144ff.

65. This is precisely what Bachem promised to the miner Peter Müller during an August Volksverein meeting. See his comments in BM (Sulzbach) to LR (Saarbrücken), 30 August 1892 (copy), in LHAK, 442/4250.

66. "Beiträge zur sozialen Fage. Mehr Fühlung mit dem Volk!" *EW,* 30 November 1890, no. 48, 380.

67. Pastor de Wyl to LR (Saarbrücken), 11 November 1890, and Pastor de Wyl to LR (Saarbrücken), 10 December 1890, in LAS, Bestand Landratsamt Saarbrücken, S/3; *SZ,* 31 July 1893, no. 177, evening edition.

68. Dieter Fricke, "Gesamtverband evangelischer Arbeitervereine Deutschlands (GEA) 1890–1933," in *Die bürgerlichen Parteien. Handbuch der Geschichte der bürgerlichen Parteien und anderer bürgerlicher Interessenorganisationen vom Vormärz bis zum Jahre 1945,* vol. 2, ed. Fricke (Berlin: Das Europäische Buch, 1968), 152–61.

69. *EW,* 29 July 1900, no. 30, 237.

70. See Hubert Nold, "Die Aufgaben der evangel. Arbeitervereine," *EW,* 14 August 1904, no. 33, 360.

71. "Arbeiterfreundlich oder arbeiterfeindlich?" *EW,* 3 November 1889, no. 44, 349–50.

72. See, e.g., "Der Rechtsschutzverein ein sozialdemokratischer Ableger," *EW,* 9 March 1890, no. 10, 74–76; and "Wohin treibt der Rechtsschutzverein?" *EW,* 15 March 1891, no. 11, 81–82.

73. Fauth, *Was ist seitens des Pfarramtes zu thun, um das weitere Umsichgreifen der Sozialdemokratie in unsern Gemeinden mit Erfolg zu bekämpfen?* (Neunkirchen: Buchdruckerei von C. A. Ohle, 1892), 26; Fauth, *Die Sozialdemokraten, was sie wollen und wie sie sind. Ein Wort der Belehrung und ernsten Mahnung an die deutschen Arbeiter* (Herborn: Buchhandlung des Nassauischen Colportagevereins, 1890).

74. Zarth, "Die evangelischen Arbeitervereine. Ein Wort zur Aufklärung," *SZ,* 19 July 1893, no. 167, evening edition; Adolf Fauth, "Was wollen unsere evange. Arbeitervereine?" *EW,* 1 December 1889, no. 48, 379–80; statutes of the association in Saarbrücken, in SAS, Bestand Alt-Saarbrücken, 1755, 25.

75. The examples are reported in the *Jahresbericht des evangelischen Arbeitervereins zu Saarbrücken, 1898,* in *EW,* 19 February 1899, no. 8, 61; *Jahresbericht des evangelischen Arbeitervereins zu Saarbrücken, 1898,* in *EW,* 11 February 1899, no. 6, 45; and *MBZ,* 28 November 1892, no. 279.

76. Fauth, *Die Sozialdemokraten,* 4, 22–23, 25–31. See also Fauth, *Was ist seitens des Pfarramtes zu thun?* 18–24.

77. Zarth, "Die evangelischen Arbeitervereine."

78. "Beiträge zur sozialen Frage. Woran liegts?" *EW,* 18 January 1891, no. 3, 18–19.

79. See, e.g., *Jahresbericht des evangelischen Arbeiter-Vereins Dudweiler für 1897,* in *EW,* 13 March 1898, no. 11, 85.

80. *EW,* 19 August 1893, no. 33, 261.

81. G. Coerper, "Revolution and Arbeiterbewegung," *EW,* 14 June 1891, no. 24, 187.

82. Ibid. *EW,* 19 August 1983, no. 33, 261.

83. Zarth, "Die evangelischen Arbeitervereine."

84. *EW,* 18 June 1893, no. 25, 194–95.

85. The description is taken from the report of the association's executive committee in *SZ,* 21 June 1884, no. 143; and *Freiherr von Stumm-Halberg und die evangelischen Geistlichen im Saargebiet,* ed. Saarbrücker evangelischen Pfarrkonferenz (Göttingen: Vandenhoeck & Ruprecht, 1896), 70, which contains the quote. See also the membership list, with occupations, in *SZ,* 15 December 1885, no. 293.

86. See the statutes in LHAK, 403/6876, 35–44.

87. *Freiherr von Stumm-Halberg und die evangelischen Geistlichen im Saargebiet,* 70–71; *SZ,* 15 February 1895, no. 39.

88. *SZ,* 4 October 1894, no. 248; *SZ,* 7 October 1894, no. 250; *SJZ,* 6 October 1894, no. 235.

89. *SZ,* 8 April 1895, no. 97.

90. *SJZ,* 25 October 1895, no. 250; LR to RP, 26 October 1895, in LHAK, 442/4371, 669–70; *SJZ,* 22 November 1895, no. 25.

91. Rudolf Saum, "Die evangelische Kirche an der Saar in den Jahrzehnten sozialer Veränderungen nach 1850," in *Die evangelische Kirche an der Saar—gestern und heute,* eds. Helmut Franz and Hans-Walter Herrmann (Saarbrücken: 1975), 240–41.

92. *SZ,* 31 July 1893, no. 177, evening edition; LR (Saarbrücken) to OP, 10 November 1895, in LHAK, 403/6875, 28.

93. Stenographische Berichte über die Verhandlungen des Reichstages, 9. Legislator periode, 3. Session, 10. Sitzung, 9 January 1895, 210–12. On the bill, see Robert Lougee, "The Anti-Revolution Bill of 1894 in Wilhelmine Germany," *Central European History* 15 (1982): 224–40.

94. Adolph Wagner, *Mein Konflikt mit dem Großindustriellen und Reichstagsabgeordneten Freiherrn v. Stumm-Halberg* (Berlin: Verlag von O. Häring, 1895).

95. "Der Professoren-Sozialismus," *SGB,* 24 March 1895, no. 12, 1; *SJZ,* 26 February 1895, no. 48.

96. Mirisch to Royal Provincial School College (Koblenz), 20 January 1896, in LHAK, 403/6875, 116; Royal Provincial School College (Koblenz) to OP, 31 January 1896, in LHAK, 403/6875, 99–101.

97. OP to RP, 21 February 1896, in LHAK, 403/6876; LR (Saarbrücken) to OP, 19 February 1896, in LHAK, 403/6875, 129–31; Stumm to OP, 11 and 30 November 1895, in LHAK, 403/6876, 115–19, 63–65.

98. Stumm to OP, 11 and 24 November 1896, in LHAK, 403/6876, 115–19, 131–36; Röchling to OP, 25 November 1896, and Vopelius to OP, 22 November 1896, in ibid., 139–40, 127–29.

99. *SJZ,* 12 November 1895, no. 265.

100. Cited in *SZ,* 28 February 1896, no. 58.

101. *Freiherr von Stumm-Halberg und die evangelishcen Geistlichen im Saarbegiet,* 12, 57–69. The public declaration (*Erklärung*), signed by all but four pastors of the Saarbrücken *Kreis,* can be found in *SZ,* 28 February 1896, no. 58.

102. Stumm to OP, 28 July 1896, in LHAK, 403/6875.

103. Stumm to OP, 24 November 1896, in LHAK, 403/6876, 134–36; Kölnische Volkszeitung, 14 July 1897, no. 515, second edition, in LHAK, 403/6876, 175.

104. See the reference in the petition in LHAK, 442/6496, 258; *SZ,* 20 April 1896, no. 108; and LR to OP, 20 May 1896, in LHAK, 403/6875, 368–69.

105. LR (Saarbrücken) to RP, 4 July 1896, in LHAK, 403/6875, 453–54; report by Dr. Scheer in LHAK, 403/6876, 473–84.

106. LR (Saarbrücken) to OP, 19 February 1896, in LHAK, 403/6875, 129–31.

107. *SZ,* 13 February 1896, no. 43.

108. LR (Saarbrücken) to RP, 27 May 1896 (copy), in LHAK, 403/6875, 391–93.

109. LR to RP, 27 May 1896 (copy), and LR to OP, 2 March 1896, in LHAK, 403/6875, 391–402, 141–42.

110. The quote is from Stumm, cited in *NSZ,* 16 February 1896, no. 46, 1.

111. See the report in *General-Anzeiger-Neue Saarbrücken Zeitung,* 2 March 1896, no. 52.

112. *SZ,* 3 March 1896, no. 62; *St. Johanner Volkszeitung* (hereafter *SJVZ*), 2 March 1896, no. 57, in LHAK, 442/6496, 57.

113. *SJZ,* 9 March 1896, no. 58; *SZ,* 9 March 1896, no. 68; LR (Saarbrücken) to RP, 11 March 1896, in LHAK, 403/6875, 193–94.

114. *SZ,* 20 April 1896, no. 108.

115. *SZ,* 9 March 1896, no. 68; *SJZ,* 9 March 1896, no. 58.

116. For the quotes from the meeting, see *SZ,* 9 March 1896, no. 68; *SJZ,* 9

March 1896, no. 58; and LR (Saarbrücken) to RP, 11 March 1896, in LHAK, 403/6875, 193–94.

117. A copy of the petition can be found in LHAK, 442/6496, 257–69. Meyer was allowed to resume his position out of fear that officials would be implicated in his resignation. See OP to Stumm, 15 August 1896, in LHAK, 403/6876, 47–50.

118. Stumm to OP, 14 May 1896, in LHAK, 403/6876, 337–47.

119. Stumm to OP, 20 May 1896, in LHAK, 403/6876, 349–52 (emphasis in original).

120. "Ein Votum der nationalliberalen Partei," *GA,* 2 March 1896, no. 52.

121. Eleanor Turk, "Holding the Line: The National Liberals and the Prussian Association Law of 1897," *German Studies Review* 2 (1979): 297–316.

122. *SJSVZ,* 28 May 1897, no. 121, in LHAK, 403/6876, 165; *SZ,* 16 December 1897, no. 343, first edition; Bellot, *Hundert Jahre,* 197.

123. *SZ,* 7 February 1898, no. 36.

124. *SJZ,* 25 April 1898, no. 95.

125. *NSZ,* 29 April 1898, no. 114, 2; *NSZ,* 2 May 1898, no. 117, 2.

126. *SZ,* 10 June 1898, no. 155, first edition.

127. *SZ,* 7 June 1898, no. 152, first edition.

128. *SJZ,* 22 June 1898, no. 143, first edition.

129. Bellot, *Hundert Jahre,* 198–99.

CHAPTER 4

1. Ludwig Lindsmayer, "Geselligkeit und Selbstbestimmung. Die Vereinskultur," in *Industriekultur an der Saar. Leben und Arbeit in einer Industrieregion 1840–1914,* ed. Richard van Dülmen (Munich: Verlag C. H. Beck, 1990), 223; *Adreßbuch für die Städte St. Johann, Saarbrücken, Malstatt-Burbach und Umgebung 1900* (St. Johann: Druck und Verlag der Saardruckerei, 1900); *Adreßbuch für die Städte St. Johann, Saarbrücken, Malstatt-Burbach und Umgebung 1905* (St. Johann: Druck und Verlag der Saardruckerei, 1905), 331–37, 468–70; Antje Fuchs, "Nach der Schicht. Zur Freizeitkultur von Hüttenarbeitern," in *Stumm in Neunkirchen. Unternehmerherrschaft und Arbeiterleben im 19. Jahrhundert. Bilder und Skizzen aus einer Industriegemeinde,* ed. Richard van Dülmen and Joachim Jacob (St. Ingbert: Röhrig Verlag, 1993), 102; Klaus-Michael Mallmann and Gerhard Bungert, "Vereine und Feste vor dem Ersten Weltkrieg," in *Dudweiler 977–1977* (Saarbrücken: Saarbrücker Zeitung Verlag, 1977), 341–48.

2. LR (Saarbrücken) to RP, 15 April 1904, in LHAK, 442/4244, 531.

3. Rohe, "Die 'verspätete' Region. Thesen und Hypothesen zur Wahlentwicklung im Ruhrgebiet," in *Probleme politischer Partizipation im Modernisierungsprozess,* ed. Peter Steinbach (Stuttgart: Klett-Cotta, 1982), 231–52.

4. Mallmann, "'Aus des Tages Last machen sie ein Kreuz des Herrn . . . ?' Bergarbeiter, Religion und sozialer Protest im Saarevier des 19. Jahrhunderts," in *Volksreligiosität in der modernen Sozialgeschichte,* ed. Wolfgang Schieder (Göttingen: Vandenhoeck & Ruprecht, 1986), 152–88; Mallmann, "Die neue Attraktivität des Himmels. Kirche, Religion und Industrielle Modernisierung," in van Dülmen,

Industriekultur an der Saar, 248–57; Eric Dorn Brose, *Christian Labor and the Politics of Frustration in Imperial Germany* (Washington, DC: Catholic University of America Press, 1985).

5. In 1910, the Christian trade union cartel listed only the following workers in occupations other than mining: 1,232 steelworkers, 1,242 construction workers, 1,423 railway workers, 106 glassworkers, 49 woodworkers, 51 telegraph workers, 20 tailors,12 textile workers, and 4 transport workers. See *Bezirkskartell der christl. Gewerkschaften an der Saar. Jahresbericht 1910* (Saabrücken: Buchdruckerei der "Saar-Post," 1910), 16.

6. Nancy Fraser, "Rethinking the Public Sphere," in *Justice Interruptus: Critical Reflections on the "Postsocialist" Condition* (New York and London: Routledge, 1997), 81.

7. See the police reports on this association and the activities of Schäfer in LAS, Depositum Amt. Illingen, 997, 144–49, 150–59, 178–80, 378–79.

8. See the membership list in LAS, Depositum Sulzbach, Fach 68, no. 5, vol. 2, unpaginated.

9. Police commissioner's report on Schäfer, 7 March 1907, in LAS, Depositum Amt. Illingen, 997, 178–80.

10. *Bürgerfreund,* 31 October 1909, in LHAK, 442/3820, 810–13.

11. See the report on the Dudweiler association's meeting of 22 October 1911 in SAD, 345, unpaginated.

12. *Bürgerfreund,* 31 October 1909, in LHAK, 442/3820, 812 (emphasis in original).

13. For the petitioning, see, e.g., the minutes of the 21 May 1908 and 30 July 1908 meetings of the Malstatt-Burbach city council in *Protokoll-Buch in den Stadtverordneten Sitzungen 1905–1908,* in SAS, Bestand Bürgermeisterei Malstatt-Burbach, 1249, unpaginated.

14. Karl Schmidt to BM Ludwig (Neunkirchen), 4 December 1901, in SAN, AI, no. 9, 657–58; report on the council meeting of 20 November 1911 in ibid., no. 10, 121–23.

15. Quoted in *Bürgerfreund,* 31 October 1909, in LHAK, 442/3820, 812.

16. Dudweiler police report of the association meeting, 13 November 1911, in SAD, 501, unpaginated. See *Bürgerfreund,* 31 October 1909, in LHAK, 442/3820, 812, for similar demands.

17. The demands can be found in the 23 October police report of the 22 October 1911 meeting of the Dudweiler *Verein* in SAD, 345, unpaginated; and in the report in *SZ,* 7 April 1908, no. 97, second edition.

18. *Bürgerfreund,* 31 October 1909, in LHAK, 442/3820, 812.

19. Report on the 25 September 1906 meeting of the local civic association in LAS, Depositum Amt. Illingen, 997.

20. Report of police commissioner on the 15 May 1908 meeting of the Dillingen *Bürgerverein* in LHAK, 442/3791, 143. The quote is from a report on the 7 February 1908 meeting of the same association in ibid., 442/3790, 395.

21. Report on the Dudweiler association's meeting of 27 November 1911 in SAD, 345, unpaginated.

22. Report of 3 December 1911 meeting in the Dudweiler mayor's office, 5 December 1911, in SAD, 345, unpaginated.

23. *Saarabien vor Gericht. Bericht über den Prozeß Hilger gegen Krämer unter Benutzung stenographischer Aufzeichnungen* (Berlin: Verlag der Buchhandlung Vorwärts, 1904), 41–42.

24. See the reports in LR (Saarlouis) to RP, 18 January 1908, and police commissioner (Dillingen) to LR (Saarlouis), 7 February 1908, in LHAK, 442/3790, 221–22, 393–94. On the incident with the mayor in 1903, see *Saarpost* (hereafter *SP*), 20 August 1903, no. 189.

25. LR (Ottweiler) to BMs of the Ottweiler district, 2 January 1907, in LAS, Depositum Amt. Illingen, 992.

26. *SJSVZ,* 30 January 1908, no. 24, second edition

27. Police commissioner's report, 7 March 1907, in LAS, Depositum Amt. Illingen, 997, 178–80.

28. *Königliche Polizei-Direktion* (Saarbrücken) to RP, 6 January 1910, in LHAK, 442/3810, 824; *Polizei-Direktion* (Saarbrücken) to RP, 31 March 1910, and LR (Ottweiler) to *Polizei-Direktion* (Saarbrücken), 24 May 1910, in LHAK, 442/3789, 269–70, 272.

29. For early reports, see LR (Saabrücken) to RP, 26 August 1901, and *Polizeiverwaltung* (St. Johann) to LR (Saarbrücken), 29 August 1902, in LAS, Landratsamt Saarbrücken S/11, unpaginated.

30. Cited in Klaus-Michael Mallmann, "'Dies Gebiet ist bis jetzt noch eine vollständige terra incognita.' Die verspätete SPD im Saarrevier," in *Richtig daheim waren wir nie. Entdeckungsreisen ins Saarrevier 1815–1955,* ed. Mallmann, Gerhard Paul, Ralph Scheck, and Reinhard Klimmt (Berlin and Bonn: Dietz, 1987), 70.

31. *Saarwacht* (hereafter *SW*), 1 August 1908, no. 178; police report on Wiebelskirchen in LHAK, 403/3793, 215–19.

32. See the membership list in LAS, Landratsamt Saarbrücken, Wa 16, unpaginated. See also Klaus-Michael Mallmann, *Die Anfänge der Bergarbeiterbewegung an der Saar (1848–1904)* (Saarbrücken: Minerva-Verlag Thinnes & Nolte, 1981), 323–24.

33. Michael Sander, "Gewerkschaftsbewegung im Montanrevier—Arbeiter und ihre Organisation an der Saar," in *Das Saarrevier zwischen Reichsgründung und Kriegsende (1871–1918),* ed. Hans-Walter Herrmann (Saarbrücken: Saarbrücker Druckerei und Verlag, 1989), 52.

34. The numbers are taken from Sander, "Gewerkschaftsbewegung im Montanrevier," 53; Peter Kiefer, *Die Organisationsbestrebungen der Saarbergleute, ihre Ursachen und Wirkungen auf dem Bereich des Saarbrücker Bergbaues und ihre Berechtigung* (Sulzbach-Saar: Peter Kiefer, 1912), 226; and *Bezirkskartell der christl. Gewerkschaften,* 16.

35. The figures are calculated from Josef Bellot, *Hundert Jahre politisches Leben an der Saar unter preussischer Herrschaft (1815–1918)* (Bonn: Ludwig Röhrscheid Verlag, 1954), 233, 236–37. For the party membership figure, see Mallmann, "'Auf dem Bürgermeisteramt ist die rote Fahne zu hissen.' Arbeiter- und Soldatenräte in der Provinz," in Mallmann et al., *Richtig daheim waren wir nie,* 91.

36. "Wie kann dem Vordringen der Sozialdemokratie im Sargebiet entgegengewirkt werden," *SP,* 15 October 1904, no. 93, first edition.

37. See the court testimony of Dr. Royer, a priest in Sulzbach, in *Saarabien*

vor Gericht, 92; *Volkswacht* (hereafter *VW*), 26 March 1908, no. 73; *VW,* 1 April 1908, no. 78; and Levenstein, *Arbeiterfrage. Mit besonderer Berücksichtigung der sozialpsychologischen Seite des modernen Großbetriebes und der psycho-physischen Einwirkungen auf die Arbeiter* (Munich: Verlag Enrst Reinhardt, 1912), 327, 334.

38. *Saarindustrie und Handel* (hereafter *SIH*), 23 October 1903, no. 43, 208; *SP,* 18 September 1903, no. 214; *SP,* 14 April 1904, no. 84.

39. LR (Saarbrücken) to RP, 7 January 1905, in LHAK, 442/4445, 3–4; RP to OP, 9 January 1905, in ibid., 5–7; *NSZ,* 2 January, no. 1, in ibid., 10; *Neunkircher Volkszeitung* (hereafter *NVZ*), 3 January 1905, no. 2, in ibid., 13, and *SW,* 5 January, no. 4; *SW,* 25 March 1905, no. 72; *SW,* 1 March 1905, no. 51.

40. RP to OP, 9 January, in LHAK, 442/4445, 5–7. See also *SW,* 10 January 1905, no. 8, and 11 January 1905, no. 9; LR (Saarbrücken) to RP, 15 April 1904, in LHAK, 442/4244, 531; and *Gewerkschaftliche Nachrichten* (hereafter *GN*), 22 June 1904, in LHAK, 442/4249, 121ff.

41. See Mining Office chairman Hilger to RP, 29 December 1903, in LHAK, 442/4244, 445; and Hilger to RP, 6 October 1903, in LHAK, 442/4244, 251–52.

42. For reports on these measures, see LR (Saarbrücken) to RP, 10 September 1912, in LHAK, 403/3793, 605–9; Hilger to Minister für Handel und Gewerbe, 12 November 1902, in LHAK, 442/4244, 391; and *SW,* 17 January 1907, no. 14. See also Hilger to RP, 24 March 1904, in LHAK, 442/4244, 479.

43. See, for activities in Dillingen, *GN,* 20 September 1904, in LHAK, 442/4249, 497; for activities in Burbach, Völklingen, and Hostenbach, *GN,* 22 June 1904, in LHAK, 442/4249, 121; and, on distributions, *GN,* 5 December 1904, in LHAK, 442/4249, 595.

44. See the report, with an accompanying blacklist of thirteen Röchling steelworkers, in *GN,* 8 April 1904, in LHAK, 442/4244, 741; the Saarbrücken *Landrat*'s report of 15 April 1904 in LHAK, 442/4244, 528; Hilger to RP, 29 May 1904, in LHAK, 4249, 51–53; and *GN,* 13 September 1904, in ibid., 373.

45. LR (Saarbrücken), 15 April 1904, in LHAK, 442/4244, 528; LR (Ottweiler) to RP, 10 September 1912, in LHAK, 402/3793, 607.

46. *VW,* 5 June 1908, no. 130; Hilger to RP, 24 March 1904, in LHAK, 442/4244, 477; *GN,* 22 June 1904, in LHAK, 442/4249, 121.

47. See the police report of 16 March 1903 on a 15 March socialist metalworkers meeting in St. Johann in LAS, Bestand Landratsamt Saarbrücken, S/11, unpaginated; BM (St. Johann) to LR (Saarbrücken), 13 April 1902, in LAS, Bestand Landratsamt Saarbrücken, S/11, unpaginated; and LR (Saarbrücken) to RP, 31 August 1903, in LHAK, 442/4244, 221. For examples of women attending SPD political meetings after 1908, see LR (Ottweiler) to RP, 19 and 24 October 1910, LHAK, 442/3789, 629–30, 625–26. The information on Frau Pflüger is contained in LR (Ottweiler) to RP, 10 September 1910, in ibid., 569.

48. On the printers' successful agitation, see the employers' report compiled by Tille for *GN* in LHAK 442/4244, 705. On the brewery workers, see LR (Saarbrücken) to RP, 15 April 1904, in LHAK, 442/4244, 529; and Tille's report for *GN* in LHAK442/4244, 733. On the bricklayers and their blacklisting, see Hilger to RP, 26 May 1904, in LHAK, 442/4249, 79–92.

49. See the official strike report compiled by the mayor of Friedrichsthal in LHAK, 442/3792, 477–78.

50. Florian Tennstedt, *Vom Proleten zum Industriearbeiter. Arbeiterbewegung und Sozialpolitik in Deutschland 1800 bit 1914* (Cologne: Bund-Verlag, 1983), 429–70.

51. *SW,* 17 June 1908, no. 139.

52. *Oberbürgermeister* (Saarbrücken) to RP, 31 January 1914, in LHAK, 442/3788, 461–62.

53. LR (Saarbrücken) to RP, 19 December 1903, in LHAK, 442/4244, 431–33; Chairman Hilger (Mining Office) to RP, 5 November 1903, in ibid., 359–60; LR (Saarbrücken) to RP, 13 November 1903, in ibid., 411–12.

54. *SZ,* 14 December 1906, no. 342, first edition; *SW,* 17 December 1906, no. 294, supplement; *SW,* 3 April 1907, no. 77; *SW,* 6 April 1907, no. 80, supplement; *SW,* 12 April 1907, no. 85, supplement; *SW,* 16 May 1907, no. 112.

55. LR (Saarbrücken) to RP, 2 February 1914, and *Oberbürgermeister* (Saarbrücken) to RP, 31 January 1914, in LHAK, 442/3788, 459–60, 462–63. On conditions in Neunkirchen, see LR (Ottweiler) to RP, 31 January 1914, in LHAK, 442/3788, 453–55.

56. See *Es werde Licht! In ernster Zeit. Ein erstes Wort an die Hüttenleute und Metallarbeiter im Saargebiet* (Saabrücken: n.p., 1906), in LHAK, 442/3792, 559–73; and Karl Alfred Gabel, *Kämpfe und Werden der Hüttenarbeiter-Organisationen an der Saar* (Saarbrücken: Saarbrücker Druckerei und Verlag, 1921), 157–60.

57. The quotes are taken from *Es werde Licht!*

58. The record is reprinted in *Saarabien vor Gericht.*

59. Cited in *Saarabien vor Gericht,* 7–8.

60. Cleff to minister of commerce, 24 November 1910, in LHAK, 442/3789, 753–85.

61. Cleff to minister of commerce, 24 November 1910, in LHAK, 442/3789, 771.

62. Cleff to minister of commerce, 24 November 1910, in LHAK, 442/3789, 767. See also LR (Ottweiler) to RP, 14 September 1908, in LHAK, 442/3791, 301–2; and LR (Ottweiler) to RP, 10 September 1910, in LHAK, 442/3789, 571.

63. Michael Sander, "Katholische Geistlichkeit und Arbeiterorganisation. 2 Gutachten aus dem Saarrevier und die Vorgeschichte des 'Fuldaer Pastorale' von 1900," in *Soziale Frage und Kirche im Saarrevier. Beiträge zu Sozialpolitik und Katholizisums im späten 19. und frühen 20. Jahrhundert* (Saarbrücken: Saarbrücker Druckerei und Verlag, 1984), 275–77; Horstwalter Heitzer, *Georg Kardinal Kopp und der Gewerkschaftsstreit* (Cologne: Böhlau, 1983).

64. "Was leistet der Verband der kathol. Arbeiter-Vereine (Sitz Berlin)?" in Bistumsarchiv Trier (hereafter BAT), 108, no. 459, 105.

65. "Leitsätze für einen Sozialen Kursus," in LHAK, 442/6220, 55–65; "Bericht über Entwicklung und Tätigkeit," in BAT, 108, no. 459, 59.

66. The figures are from Michael Sander, "Katholische Arbeitervereine Berliner Richtung," *Archiv für mittelrheinische Kirchengeschichte* 37 (1985): 131; and Kiefer, *Die Organisationsbestrebungen der Saarbergleute,* 90.

67. *SJSVZ,* 11 November 1904, no. 260, first edition; "Bericht über Entwicklung und Tätigkeit des Verbandes der kathol. Arbeitervereine der Diözese Trier

und über dessen 2. Delegiertentag am 31. Juli und 1. August in Neunkirchen, Bez. Trier," in BAT, 108, no. 459, 71–86; LR (Saarbrücken) to RP, 24 October 1903, in LAS, Landratsamt Saarbrücken, S/11, unpaginated.

68. "Leitsätze für einen Sozialen Kursus," in LHAK, 442/6220, 55, 56.

69. *SJSVZ,* 16 August 1904, no. 245, first edition.

70. The quotes are taken from a report in *SJSVZ,* 19 July 1905, no. 162, second edition; the statutes of the Catholic workers' association in Lautzkirchen, in LAS, Nachlass Jakob Kraus, no. 7, unpaginated; and an 1884 description of Catholic workers' associations in *Arbeiterwohl,* cited in Sander, "Katholische Geistlichkeit und Arbeiterorganisation," 274.

71. See *Feierabend. Unterhaltungs-Beilage zum Arbeiter,* 2 February 1902, no. 4, in BAT, 108, no. 459, 49; "Ordnung hilft Haushalt," *Feierabend,* 22 January 1905, no. 4; and the series entitled "Feinde der Sparsamkeit sind vor allem Trägheit, Leichtsinn und Verschwendung" beginning in *Feierabend,* 12 March 1905, no. 11.

72. *GN,* 9 February 1905, in LHAK, 442/4445, 184–89; *SJSVZ,* 30 November 1904, no. 275, first edition; *SJSVZ,* 3 August 1904, no. 175.

73. Cited in *GN,* 9 February 1905, in LHAK, 442/4445, 485, 487.

74. Cited in *GN,* 16 November 1909, in LHAK, 442/3820, 780–82.

75. Michael Schneider, *Die Christliche Gewerkschaften 1894–1933* (Bonn: Verlag Neue Gesellschaft, 1982).

76. Sander, "Gewerkschaftsbewegung im Montanrevier," 47; Peter Kiefer, *25 Jahre Gewerkvereine christlicher Bergarbeiter im Saarrevier* (Saarbrücken: Verlag des Gewerkvereins christlicher Bergarbeiter, Deutschlands, 1929), 23–24.

77. *Berzirkskartell der christl. Gewerkschaften,* 11–12.

78. See the *Gewerkverein* handbill entitled "Ein Wort zur Aufklärung und Beherzigung" in LHAK, 442/3792, 391; and Horstwalter Heitzer, "Die christliche Bergarbeiterbewegung im Saarrevier von 1904 bis zum Ersten Weltkrieg," in *Soziale Frage und Kirche im Saarrevier,* 253–54.

79. Mining Office chairman Cleff to minister of commerce, 24 November 1910, in LHAK, 442/3789, 773, 779.

80. "Ein Wort zur Aufklärung und Beherzigung," in LHAK, 442/3792, 589–91.

81. *Protokoll der 12. General-Versammlung des Gewerkvereins christlicher Bergarbeiter in der städtischer Tonhalle zu Saarbrücken vom 27. bis 30. Juni 1909 einschliesslich,* in Stadtbibliothek Mönchengladbach (hereafter SBMG), 6 Iwa/9, 217; *Geschäftsbericht des Vorstandes des Gewerkvereins christlicher Bergarbeiter Deutschland für die Jahre 1907 und 1908* (Essen: Verlag des Gewerkverein christlicher Bergarbeiter Deutschlands, 1908), 91; Heitzer, "Die christliche Bergarbeiterbewegung," 249; *SP,* 11 October 1906, no. 235.

82. Sulzbach police report, 13 November 1907, LAS, Depositum Sulzbach, 68, no. 2; *Protokoll der 12. General-Versammlung des Gewerkvereins christlicher Bergarbeiter,* in SBMG, 6 Iwa/9, 205, 217.

83. *Protokoll der 12. General-Versammlung des Gewerkvereins christlicher Bergarbeiter,* SBMG, 6 Iwa/9, 589 90.

84. Gerhard Bungert and Klaus-Michael Mallmann, "Burbach 1906: Der erste Hüttenstreik an der Saar," *Arbeitnehmer* 7–8 (1977): 337–40.

85. *SP,* 2 April 1906, no. 76, first edition; *SP,* 4 April 1906, no. 78.

86. The quotes are from the police report on a 25 May 1906 meeting chaired by Wernerus and held in the Saalbau in Saarbrücken, LHAK, 442/3792, 458–59.

87. *SP,* 24 March 1906, no. 69, first edition; *SP,* 7 April 1906, no. 81, first edition.

88. *SP,* 9 June 1906, no. 131, first edition.

89. Schneider, *Die Christlichen Gewerkschaften,* 196, 245; Brose, *Christian Labor,* 192.

90. *Bezirkskartell der christl. Gewerkschaften,* 14.

91. For this policy, see Chairman Cleff to minister of commerce, 24 November 1910, in LHAK, 442/3789, 753–85; LR (Saarbrücken) to RP, 6 December 1910, in LHAK, 442/3785, 19; and *SP,* 23 June 1905, no. 141, in LHAK, 442/4445, 425.

92. Cited in Klaus Saul, *Staat, Industrie, Arbeiterbewegung. Zur Innen- und Aussenpolitik des Wilhelminischen Deutschland 1903–1914* (Düsseldorf: Bertelsmann Universitätsverlag, 1974), 185; and Michael Sander, "Zwichen Kirche, Streik, und Zentrum. Der Gewerkverein christlicher Bergarbeiter im Krisenjahr 1912," in Mallmann et al., *Richtig daheim waren wir nie,* 90.

93. See the firm's notice in LAS, Landratsamt Saarbrücken, S/12, unpaginated.

94. Heitzer, "Die christliche Bergarbeiterbewegung," 251–59.

95. Dechant Backes, *Memorandum zur Bergarbeiterstreikbewegung im Saarrevier 1912–13* (Neunkirchen: Verlag der "Neunkirchener Zeitung," 1913), 41; Heinrich Imbusch, *Die Saarbergarbeiter-Bewegung 1912/13* (Cologne: Christl. Gewerkschaftsverlag, 1913).

96. LR (Ottweiler) in LHAK, 442/3790, 291–309. The quotes are from pp. 293 and 297.

97. See the list of councillors in the municipal council in *Adreßbuch von Neunkirchen 1910/11* (Neunkirchen: Buchdruckerei C. A. Ohle, 1911), 4, in SAN, I, Verwaltung Bereich 1; and the report on a Center Party meeting in 1909 in LHAK, 442/3820, 850–52.

98. *SP,* 13 December 1912, no. 285, in LHAK, 442/3785, 651.

99. Bellot, *Hundert Jahre,* 210–13, 221–24, 233–36.

100. *SZ,* 26 October 1903, no. 294. For these developments at the national level, see Holger J. Tober, *Deutscher Liberalismus und Sozialpolitik in der Ära des Wilhelminismus. Anschauungen der liberalen Parteien im parlamentarischen Entscheidungsprozess und in der öffentlichen Diskussion* (Husum: Matthiesen Verlag, 1999); Theodor Eschenburg, *Das Kaiserreich am Scheideweg. Bassermann, Bülow und der Block* (Berlin: Verlag für Kulturpolitik, 1929); and Beverly Heckart, *From Bassermann to Bebel: The Grand Bloc's Quest for Reform in the Kaiserreich, 1900–1914* (New Haven: Yale University Press, 1974).

101. Bellot, *Hundert Jahre,* 208–10.

102. See "Was will der 'Jungliberale Verein'?" excerpted from a lecture delivered by the association's chairman, the rector P. Venter, on 8 June 1907, in LAS, Depositum Sulzbach, Fach 68, no. 2, unpaginated.

103. See "Satzungen des Jungliberalen Vereins Hühnerfeld" and the membership list in LAS, Depositum Sulzbach, Fach 68, no. 3, unpaginated. Figures for the Saarbrücken association are in *NSZ,* 3 May 1904, no. 103, second edition, 1.

104. *NSZ,* 6 July 1903, no. 156, 2; *NSZ,* 29 July 1903, no. 176, 2; *NSZ,* 10 August 1903, no. 186, 2; *SZ,* 13 October 1906, no. 281, second edition

105. See the circulars from Heinrich Boltz and Theodor Vogel of the National Liberal executive committee in Saarbrücken to the trustees of the local electoral associations in the district, dated 17 December 1906, 22 December 1906, and 5 January 1907, in LAS, Landratsamt Saarbrücken, Wa 19, unpaginated. The quotes are from the circular of 5 January (emphasis in original).

106. The quote is from the report by Krümmer (Mining Office) to RP, 25 February 1907, in LHAK, 442/3810, 111–35.

107. Bellot, *Hundert Jahre,* 221ff.

108. *NSZ,* 14 January 1907, no. 11; *NSZ,* 4 February 1907, no. 29; *NSZ,* 4 December 1911, no. 284; *GA,* 18 December 1911, no. 296; *GA,* 9 January 1912, no. 7.

109. Quoted in *VW,* 28 May 1908, no. 124.

CHAPTER 5

1. Heidrun Homburg, "Anfänge des Taylorsystems in Deutschland vor dem Ersten Weltkrieg. Eine Problemskizze unter besonderer Berücksichtigung der Arbeitskämpfe bei Bosch 1913," *Geschichte und Gesellschaft* 4 (1978): 171.

2. Harry Braverman, *Labor and Monopoly Capital: The Degradation of Work in the Twentieth Century* (New York: Monthly Review Press, 1998), 59–85; David F. Noble, *America By Design: Science, Technology, and the Rise of Corporate Capitalism* (Oxford: Oxford University Press, 1977), 264–77; Reinhard Bendix, *Work and Authority in Industry: Ideologies of Management in the Course of Industrialization* (New York: Harper & Row, 1956), 274–81; Robert A. Brady, *The Rationalization Movement in German Industry: A Study in the Evolution of Economic Planning* (Berkeley: University of California Press, 1933), 33–51.

3. Alfred D. Chandler, *Scale and Scope: The Dynamics of Industrial Capitalism* (Cambridge, MA: Harvard University Press, 1990), 393–502.

4. Wilfried Spohn, "Religion and Working-Class Formation in Imperial Germany, 1871–1914," *Politics and Society* 19 (1991): 109–32.

5. Richard Woldt, *Der industrielle Großbetrieb. Eine Einführung in die Organisation moderner Fabrikbetriebe* (Stuttgart: J. H. W. Dietz, 1911), 5.

6. Jürgen Kocka, "Industrielles Management: Konzeptionen und Modelle in Deutschland vor 1914," *Vierteljahrschrift für Sozial- und Wirtschaftsgeschichte* 56 (1969): 356–60; Kocka, *Unternehmer in der detuschen Industrialisierung* (Göttingen: Vandenhoeck & Ruprecht, 1975), 105–15.

7. This assessment agrees with Woldt's in *Der industrielle Grossbetrieb* and Kocka's in "Industrielles Management." The reference to "antitheoretical" paternalism comes from "Industrielles Management," 335.

8. Christoph Lang, "'Herren im Hause.' Die Unternehmer," in *Industriekultur an der Saar. Leben und Arbeit in einer Industrieregion 1840–1914,* ed. Richard van Dülmen (Munich: C. H. Beck, 1989), 136–37; Fritz Hellwig, "Unternehmer und Unternehmungsform im saarländischen Industriegebiet," *Jahrbuch für Nationalökonomie und Statistik* 158 (1943): 421–22.

9. Hellwig, "Unternehmer und Unternehmungsform," 424.

10. The comparative figures for the Röchling steelworks are in Alexander Tille, *Das Haus Röchling und seine Unternehmen* [Saarbrücken: n.p., 1907], 256.

11. Richard Woldt, *Das großindustrielle Beamtentum. Eine gewerkschaftliche Studie* (Stuttgart: J. H. W. Dietz, 1911), 8–69; Otto Neuloh, *Die Deutsche Betriebsverfassung und ihre Sozialformen bis zur Mitbestimmung* (Tübingen: J. C. B. Mohr [Paul Siebeck], 1956), 138–48.

12. Woldt, *Der industrielle Großbetrieb,* 80–83; Dieter Groh, "Intensification of Work and Industrial Conflict in Germany," *Politics and Society* 9 (1978): 361–62.

13. Gunnar Stollberg, *Die Rationalisierungsdebatte 1908–1933. Freie Gewerkschaften zwischen Mitwirkung und Gegenwehr* (Frankfurt and New York: Campus Verlag, 1981), 32–36; Thomas Welskopp, *Arbeit und Macht im Hüttenwerk. Arbeits- und industrielle Beziehungen in der deutschen und amerikanischen Eisen- und Stahlindustrie von den 1860er bis zu den 1930er Jahren* (Bonn: J. H. W. Dietz, 1994), 535–59.

14. Gerhard Ames, "Arbeiten auf der Hütte. Lebenskontexte von Hüttenarbeitern," in van Dülmen, *Industriekultur an der Saar,* 116. For more general discussions, see Hans Ehrenberg, *Die Eisenhüttentechnik und der deutsche Hüttenarbeiter* (Stuttgart and Berlin: J. G. Cotta, 1906), 142ff.; and Gerhard A. Ritter and Klaus Tenfelde, *Arbeiter im Deutschen Kaiserreich 1871 bis 1914* (Bonn: J. H. W. Dietz, 1992), 357–59.

15. Welskopp, *Arbeit und Macht,* 372–73; Ames, "Arbeiten auf der Hütte," 116.

16. See the tables, compiled by Herrmann Röchling, included in *Königlicher Gewerbeinspektor* (Saarbrücken) to RP, 5 July 1909, in LHAK, 442/8102, 141–43, 145–47, 149–51, 153–57, 159–61, 183, 185–87. See also Hans-Walter Herrmann, "Die Verstetigung der Kraft. Von der Werkstatt zur Fabrik," in van Dülmen, *Industriekultur an der Saar,* 78; and Ames, "Arbeiten auf der Hütte," 115–16.

17. Welskopp, *Arbeit und Macht,* 437–56. The quote is from p. 437.

18. *Die Burbacher Hütte 1856–1906. Denkschrift zur Feier des fünfzigjährigen Bestehens der Hütte am 22. Juni 1906* (Saarlouis: Franz Stein Nachf. Hausen & Co., 1906), 29, 65.

19. Richard Nutzinger, *Karl Röchling. Das Lebenswerk eines Großindustriellen* (Völklingen-Saarbrücken: Buch- und Kustdruckerei Gebr. Hofer A.-G., 1927), 112–13.

20. *Die Burbacher Hütte,* 29, 65. See also Julius Kollmann, *Die Großindustrie des Saargebiets. Eine zusammenhängende Darstellung der geschichtlichen und technischen Entwicklung bis auf den gegenwärtiger Stand* (Stuttgart: Verlag der Technischen Monatshefte, Franckh'sche Verlagshandlung, 1911), 24–66; Herrmann, "Die Verstetigung der Kraft," 75–82; and Harald Glaser, "Thomasstahl und Tafelglas. Die Eisen- und Glashütten zwischen 1840 und 1918," in van Dülmen, *Industriekultur an der Saar,* 65–71 (for a discussion of similar trends in the Saar glass industry).

21. Anson Rabinbach, *The Human Motor: Energy, Fatigue, and the Origins of Modernity* (Berkeley and Los Angeles: University of California Press, 1990), 189–202; Joan Campbell, *Joy in Work, German Work: The National Debate, 1800–1945* (Princeton, NJ: Princeton University Press, 1989), 73–106.

22. Rabinbach, *Human Motor,* 4, 189–202.

23. On the need for the "rational cultivation of labor power," see especially Adolf Weber, *Der Kampf zwischen Kapital und Arbeit. Versuch einer systematischen Darstellung mit besonderer Berücksichtigung der gegenwärtigen deutschen Verhältnisse* (Tübingen: J. C. B. Mohr [Paul Siebeck], 1910), 93ff.

24. Other members of this group of critics of *Sozialpolitik,* in large part organized around the *ZfS,* included Andreas Voigt, Adolf Weber, Ludwig Bernhard, Bernhard Harms, Paul Moldenhauer, and Fritz Stier-Somlo. See Rüdiger vom Bruch, *Wissenschaft, Politik und öffentliche Meinung. Gelehrtenpolitik im Wilhelminischen Deutschland (1890–1914)* (Husum: Matthiesen, 1980), 297ff.; and Dieter Lindenlaub, "Firmengeschichte und Sozialpolitik. Zur Wissenschaftsgeschichte der Nationalökonomie in Deutschland vor dem Ersten Weltkrieg," in *Wissenschaft Wirtschaft und Technik. Studien zur Geschichte. Wilhelm Treue zum 60. Geburtstag,* ed. Karl-Heinz Manegold (Munich: Bruckmann, 1969), 272–85.

25. Hartmut Kaelble, *Industrielle Interessenpolitik in der Wilhelminischen Gesellschaft. Centralverband Deutscher Industrieller 1895–1914* (Berlin: Walter de Gruyter & Co., 1967), 185–89; Klaus Saul, *Staat, Industrie, Arbeiterbewegung. Zur Innen- und Aussenpolitik des Wilhelminischen Deutschland 1903–1914* (Düsseldorf: Bertelsmann Verlag, 1974), 363–64; Dirk Stegmann, *Die Erben Bismarcks. Parteien und Verbände in der Spätphase des Wilhelminischen Deutschlands. Sammlungspolitik* 1897–1918 (Cologne and Berlin: Kiepenheuer & Witsch, 1970), 236–37.

26. Arnim Tille, *Ein Kämpferleben. Alexander Tille 1866–1912* (Gotha: Friedrich Andreas Perthes A-G, 1916), 37, 85–87; Fritz Hellwig, *Die Saarwirtschaft und ihre Organisationen seit der Errichtung der Industrie- und Handelskammer zu Saarbrücken 1863/64* (Saarbrücken: Buchgewerbehaus Aktiengesellschaft Saarbrücken, 1939), 80–81, 85; Wilfried Schungel, *Alexander Tille (1866–1912). Leben und Ideen eines Sozialdarwinisten* (Husum: Matthiesen Verlag, 1980), 10–13; Kaelble, *Industrielle Interessenpolitik,* 188.

27. Tille, "Ein wirtschaftliches Einlenken der nationallieralen Partei?" in *Südwestdeutsche Wirtschaftszeitung* (hereafter *SWDWZ*), 30 April 1909, no. 18, 203–4.

28. See Wolf's remarks cited in vom Bruch, *Wissenschaft, Politik, und öffentliche Meinung,* 306; and Ludwig Pohle, *Die gegenwärtige Krisis in der deutschen Volkswirtschaftslehre* (Leipzig: A. Deichert'sche Verlagsbuchhandlung Nachf., 1911), v.

29. Julius Wolf, *Nationalökonomie als exakte Wissenschaft. Ein Grundriss* (Leipzig: A. Deichert'sche Verlagsbuchhandlung Nachf., 1908), 19–20.

30. Ibid., 21–22; Wolf, *Die Volkswirtschaft der Gegenwart und Zukunft. Die wichtigsten Wahrheiten der allgemeinen Nationalökonomie* (Leipzig: A. Deichert'sche Verlagsbuchhandlung Nachf., 1912), 44–48.

31. Ehrenberg, "Bisherige Hauptergebnisse und nächste Aufgaben der exakt-vergleichenden Wirtschaftsforschung (1910)," in *Zur Einführung in die exakt-vergleichende Wirtschaftsforschung. Aufsätze aus dem Archiv für exakte Wirtschaftsforschung (Thünen-Archive* (Jena: G. Fischer, 1912), 90–91; Lindenlaub, "Firmengeschichte und Sozialpolitik," 275–76.

32. Tille, *Die Berufsstandspolitik des Handel- und Gewerbestandes,* vol. 1, *Die gewerbliche Ertragswirtschaft* (Berlin: Rosenbaum & Hart, 1910), 100.

33. Ibid., 1:107; Tille, *Die Berufsstandspolitik des Handel- und Gewerbestandes,* vol. 2, *Der Geisteskampf gegen die gewerbliche Ertragswirtschaft* (Berlin: Rosenbaum & Hart, 1910), 245—-46.

34. Tille, "Konjunktur, Leistungsunterschied und Ständebildung," *Neue Saarbrücker Gewerbeblatt* (hereafter *NSGB*), 20 July 1901, no. 29, 114–16. For examples, see *Jahresbericht der Handelskammer zu Saarbrücken für 1906* (Saarbrücken: Buchdruckerei der Neuen Saarbrücker Zeitung, 1907), I.4–I.5; *Jahresbericht der Handelskammer zu Saarbrücken für 1907* (Saarbrücken: Buchdruckerei der Neuen Saarbrücker Zeitung, 1908), I.4, I.10; and *Jahresbericht der Handelskammer zu Saarbrücken für 1910* (Saarbrücken: Buchdruckerei der Neuen Saarbrücker Zeitung, 1911), I.3, I.4.

35. *Jahresbericht der Handelskammer zu Saarbrücken für 1910,* I.4; Tille, *Das Haus Röchling,* 98.

36. Tille, *Die Berufsstandspolitik,* 1:104–5, 107; 2:245–46.

37. Tille, *Das Haus Röchling,* 255.

38. "Gutachten des Wirtschaftlichen Vereins über vorgeschlagene Herabsetzung der Arbeitszeit erwachsener Arbeiterinnen," *SIH,* 3 November 1905, no. 44, 225–26.

39. Alexander Tille, ed., *Musterarbeitsordnung der Saarindustrie,* no. 25, *Südwestdeutsche Flugschriften* (Saarbrücken: Kommissionsverlag der Buchhandlung C. Schmidtke, 1913), 22.

40. Ibid., 31.

41. Ibid., 26.

42. See the reference to this discussion in *SWDWZ,* 5 June 1908, no. 23, 146.

43. On the development of racial hygiene and German nationalism, see Hans-Walter Schmuhl, *Rassenhygiene, Nationalsozialismus, Euthanasie* (Göttingen: Vandenhoeck & Ruprecht, 1987); Jochen-Christoph Kaiser, Kurt Nowak, and Michael Schwartz, eds., *Eugenik, Sterlisation, "Euthanasie." Politische Biologie in Deutschland 1895–1914* (Berlin: Bundverlag Union, 1992), 1–50; and Paul Weindling, *Health, Race, and German Politics between National Unification and Nazism, 1870–1945* (Cambridge: Cambridge University Press, 1989).

44. Wolf, *Nationalökonomie als exakte Wissenschaft,* 22; Wolf, *Die Volkswirtschaft,* 52.

45. Tille, "Die Menschenrechte," *SWDWZ,* 3 May 1907, no. 18, 124; Tille, "Konjunktur, Leistungsunterschied und Ständebildung"; Tille, *Die Berufsstandspolitik,* 2:126.

46. Tille, *Die Berufsstandspolitik,* 1:17; Tille, "Die Menschenrechte," 124–25; Schungel, *Alexander Tille,* 36–41.

47. Tille, *Die Berufsstandspolitik,* 1:44ff.

48. Wolf, "Das Rassenproblem in der Weltwirtschaft," *Zeitschrift für Sozialwissenschaft* 6 (1903)," 30, 35–36; Wolf, *Nationalökonomie als exakte Wissenschaft,* 198.

49. Tille, *Der Wettbewerb weisser und gelber Arbeit in der industriellen Produktion* (Berlin: Otto Elsner, 1904), 5; Tille, *Von Darwin bis Nietzsche. Ein Buch Entwicklungsethik* (Leipzig: C. G. Naumann, 1895), 239–41. See also the work of economist Andreas Voigt in *Soziale Utopien. Fünf Vorträge* (Leipzig: G. J. Göschen'sche Verlagshandlung, 1906), especially 46–47.

50. Tille, *Der Wettbewerb.*

51. Wolf, "Das Rassenproblem," 30–42; Wolf, *Die Volkswirtschaft,* 49–53; Tille, *Der Wettbewerb,* 7–8, 24–26. For a general discussion, see Michael Adas, *Machines as the Measure of Men: Science, Technology, and Ideologies of Western Dominance* (Ithaca, NY: Cornell University Press, 1989), especially 306–18.

52. The essay contest, sponsored by the Ruhr industrialist Friedrich Alfred Krupp, was organized around the question "What can we learn from the theory of evolution about internal political development and state legislation?" See Sheila Weiss, *Race Hygiene and National Efficiency: The Eugenics of Wilhelm Schallmayer* (Berkeley and Los Angeles: University of California Press, 1987), 64–89. On Hugenberg, see Dankwart Guratzsch, *Macht durch Organisation. Die Grundlegung des Hugenbergschen Presseimperiums* (Düsseldorf: Bertelsmann Universitätsverlag, 1974), 89ff. On racism in the heavy industry press, see Dirk Stegmann, *Die Erben,* 282, especially n. 165.

53. Tille, *Die Berufsstandspolitik,* 2:173–74.

54. Ibid., 2:190–91; Tille, *Der Soziale Ultramontanismus und seine "katholische Arbeitervereine"* (Berlin: Otto Elsner, 1905), 38–39.

55. Tille, "Konjunktur, Leistungsunterschied und Ständebildung"; Tille, "Auf Vorposten, dem Feinde nah," *SIH,* 21 July 1905, no. 29, 149; Tille, *Die Berufsstandspolitik,* 2:191. See also the reports in the Saar employer press, including "Die Ungleichheit der Menschen in ihrer Leistungsfähigkeit," *SIH,* 18 May 1906, no. 20, 125–26; and "Die wirtschaftlich schwachen," *SWDWZ,* 14 April 1911, no. 15, 123–24.

56. Tille, *Die Berufsstandspolitik,* 2:192.

57. The quotes are from Tille, "'Coelibat im Volksdienst,"cited in Schungel, *Alexander Tille,* 54. See also the discussion in Tille, *Die Berufsstandspolitik,* 1:12–14.

58. Stegmann, *Die Erben,* especially 305ff.; Klaus Saul, *Staat, Industrie, Arbeiterbewegung,* 368ff; Steinmetz, *Regulating the Social: The Welfare State and Local Politics in Imperial Germany* (Princeton, NJ: Princeton University Press, 1993), 126, 141.

59. "Sozialmoralismus und Sozialwirtschaft," *SIH,* 15 January 1904, no. 3, 13–14.

60. Tille, "Ostlondon als Nationalheilstatt," in *Die Zukunft,* vol. 5 (1893), cited in Fritz Bolle, "Darwinismus und Zeitgeist," *Zeitschrift für Religions- und Geistesgeschichte* 14 (1962): 169; and Wilfried Schungel, *Alexander Tille,* 48. See also Tille, *Der Soziale Ultramontanismus,* 44–51; and Tille, *Lujo Brentano und der akademische Klassenmoralismus* (Berlin: Otto Elsner, 1912).

61. "Die sittlichen Schäden der Klassenversicherung des Reiches," *SWDWZ,* 13 May 1910, no. 19, 129–30.

62. Weiss, *Race Hygiene and National Efficiency,* 85.

63. Tille called for the replacement of the Christian and democratic "morality of sympathy" with the scientific "morality of genus" (*Gattungsmoral*), which recognized evolution and the process of natural selection as the highest moral ideal. See Tille, *Von Darwin bis Nietzsche,* 24, 111–13.

64. See, e.g., the miners' flyer entitled "Ein Wort zur Aufklärung und Beherzigung an die christlich-nationale Arbeiterschaft des Saarreviers," delineating their reform program of 1905–6, in LHAK, 442/3792, 590ff.

65. "Das Ende der Wohlfahrtseinrichtungen," *SWDWZ,* 19 June 1908, no. 25, 155–56.

66. Tille, *Die Berufsstandspolitik,* 1:12.

67. "Der ursprüngliche Sinn der deutschen Klassenversicherung," *SWDWZ,* 24 June 1910, no. 25, 164; Tille, *Die Berufsstandspolitik,* 1:263–64.

68. *NSGB,* 18 September 1903, no. 38, 177; minutes of the Saar Chamber of Commerce meeting of 24 November 1903 and questionnaire in *SIH,* 4 December 1903, no. 49, 241–42; *SIH,* 10 July 1903, no. 28, 137; "Zur Ständebildung in der industriellen Handarbeiterschaft," *SIH,* 20 October 1905, no. 42; *SIH,* 1 June 1906, no. 22, 136–37.

69. "Das Ende der Wohlfahrtseinrichtungen."

70. Kathleen Canning, "Social Policy, Body Politics: Recasting the Social Question in Germany, 1875–1900," in *Gender and Class in Modern Europe,* ed. Laura L. Frader and Sonya O. Rose (Ithaca, NY: Cornell University Press, 1996), 211–37; Carola Sachse, *Siemens, der Nationalsozialismus und die moderne Familie. Eine Untersuchung zur sozialen Rationaliserung in Deutschland im 20. Jahrhundert* (Hamburg: Rasch und Röhring Verlag, 1990); Alf Lüdtke, "Arbeiterpolitik versus Politik der Arbeiter? Zu Unternehmensstrategien und Arbeiterverhalten in deutschen Großbetrieben zwischen 1890 und 1914/20," in *Arbeiter und Bürger im 19. Jahrhundert,* ed. Jürgen Kocka (Munich, 1986), 202–12.

71. The quote is from Tille, *Das Haus Röchling,* 255. For discussions of this wider biopolitical interest, see Weindling, *Health, Race, and German Politics,* 214ff.; Geoff Eley, "German History and the Contradictions of Modernity: The Bourgeoisie, the State, and the Mastery of Reform," in *Society, Culture, and the State in Germany,* ed. Eley (Ann Arbor: University of Michigan Press, 1996); and Edward Ross Dickinson, "Biopolitics, Fascism, and Democracy: Some Reflections on Our Discourse about 'Modernity,'" *Central European History* 37 (2004): 1–48.

72. Röchling'sche Eisen- und Stahlwerke, GmbH, Völklingen an der Saar, *Studien und soziale Aufgaben sowie deren Lösung* (n.p., 1906), 13.

73. "Burbacher Hütte bei Saarbrücken. Wohlfahrtseinrichtungen 1897," in SAS, Bestand Bürgermeisterei Malstatt-Burbach, file 379, 11–14; *Die Burbacher Hütte,* 114; Richard Nutzinger, Hans Boehmer and Otto Johannsen, eds., *50 Jahre Röchling Völklingen. Die Entwicklung eines rheinischen Industrie-Unternehmens* (Saarbrücken-Völklingen: Hofer, 1931), 40–45.

74. *Die Burbacher Hütte,* 103.

75. Tille, *Das Haus Röchling,* 287.

76. Ibid., 289; Wilhelm Kulemann, *Die Berufsvereine. Geschichtliche Entwicklung der Berufsorganisationen der Arbeitnehmer und Arbeitgeber aller Länder,* vol. 2 (Jena: G. Fischer, 1908), 466; Karl Alfred Gabel, *Kämpfe und Werden der Hüttenarbeiter-Organisationen an der Saar* (Saarbrücken: Saarbrücker Druckerei und Verlag, 1921), 164.

77. *Rede des Herrn Langerichtsdirektors Dr. Karl Röchling, Mitglied des Abgeordnetenhauses, über Die Werkvereinsbewegung. Gehalten bei Gelegenheit der Fahnenweihe des Hüttenarbeitervereins E. V. Völklingen und des 25 jährigen Stiftungsfestes des Hüttenvereins "Stahl u. Eisen" zu Völklingen am 18. Juli 1909* (Völklingen:

C. H. Scheuer, [1909], 7; Saarbrücken *Landrat*'s report on labor relations from 6 December 1910 in LHAK, 442/3785, 22.

78. Tille, *Das Haus Röchling,* 255.

79. Röchling'sche Eisen- und Stahlwerke, *Studien und soziale Aufgaben sowie deren Lösung,* 13.

80. Ibid., 12; Tille, *Das Haus Röchling,* 281–82.

81. Tille, *Das Haus Röchling,* 284; *Die Burbacher Hütte,* 104–5.

82. "Burbacher Hütte bei Saarbrücken. Wohlfahrtseinrichtungen 1897," 24; Tille, *Das Haus Röchling,* 284.

83. *SZ,* 15 October 1906, no. 183, first edition.

84. Tille, *Das Haus Röchling,* 284,

85. Röchling'sche Eisen- und Stahlwerke, *Studien und soziale Aufgaben sowie deren Lösung,* 5.

86. Ibid., 2, 5–6; Tille, *Das Haus Röchling,* 277–78.

CHAPTER 6

1. See Fritz Stern, *The Politics of Cultural Despair* (Berkeley and Los Angeles: University of California Press, 1961), xvi–xvii, 58, 230–31, 258; George Mosse, *The Crisis of German Ideology* (New York: Universal Library, 1964), 13ff., 282–91, 312–16; and, much more circumspect and still highly illuminating, Ralph E. Bowen, *German Theories of the Corporative State* (New York and London: Whittlesey House, 1947), 9.

2. Hans-Jürgen Puhle, "Parlament, Parteien und Interessenverbände 1890–1914," in *Das Kaiserliche Deutschland. Politik und Gesellschaft 1870–1918,* ed. Michael Stürmer (Düsseldorf: Droste Verlag, 1970), especially 341, 361, 364; Thomas Nipperdey, "Interessenverbände und Parteien in Deutschland vor dem Ersten Weltkrieg," in *Moderne deutsche Sozialgeschichte,* ed. Hans-Ulrich Wehler (Cologne and Berlin: Kiepenheuer & Witsch, 1966), especially 376, 387; Hans-Peter Ullmann, *Interessenverbände in Deutschland* (Frankfurt am Main: Suhrkamp, 1988), 121–22.

3. Jürgen Kocka, "Vorindustrielle Faktoren in der deutschen Industrialisierung. Industriebürokratie und 'neuer Mittelstand,'" in Stürmer, *Das kaiserliche Deutschland,* 278–79; Hans-Ulrich Wehler, *The German Empire, 1871–1918* (Leamington Spa: Berg Publishers, 1985), 70–71; Heinrich August Winkler, "Unternehmerverbände zwischen Ständeideologie und Nationalsozialismus," *Vierteljahrshefte für Zeitgeschichte* 17 (1969): 369. See also Mary Nolan, *Visions of Modernity: American Business and the Modernization of Germany* (New York and Oxford: Oxford University Press, 1994), 190, 192.

4. Philippe C. Schmitter and Gerhard Lehmbruch, eds., *Trends Toward Corporatist Intermediation* (London: Sage Publications, 1979); Suzanne Berger, ed., *Organizing Interests in Western Europe: Pluralism, Corporatism, and the Transformation of Politics* (Cambridge: Cambridge University Press, 1981); Charles Maier, *Recasting Bourgeois Europe: Stabilization in France, Germany, and Italy in the Decades after World War One* (Princeton, NJ: Princeton University Press, 1975);

Heinrich August Winkler, ed., *Organisierter Kapitalismus* (Göttingen: Vandenhoeck & Ruprecht, 1974).

5. Philippe C. Schmitter, "Still the Century of Corporatism?" in Schmitter and Lehmbruch, *Trends Toward Corporatist Intermediation,* 21–22.

6. Claus Offe, "The Attribution of Public Status to Interest Groups: Observations on the West German Case," and Philippe C. Schmitter, "Interest Intermediation and Regime Governability in Contemporary Western Europe and North America," in Berger, *Organizing Interests in Western Europe,* 136, 294–95.

7. Matthew H. Elbow, *French Corporative Theory, 1789–1948: A Chapter in the History of Ideas* (New York: Columbia University Press, 1953), especially 69–75; David D. Roberts, *The Syndicalist Tradition and Italian Fascism* (Chapel Hill: University of North Carolina Press, 1979).

8. Bowen, *German Theories of the Corporative State.*

9. Klaus Saul, *Staat, Industrie, Arbeiterbewegung im Kaiserreich. Zur Innen- und Aussenpolitik des Wilhelminischen Deutschland 1903–1914* (Düsseldorf: Bertelsmann Universitätsverlag, 1974), 98–115; Hans-Peter Ullmann, "Unternehmerschaft, Arbeitgeberverbände und Streikbewegung 1890–1914," in *Streik. Zur Geschichte des Arbeitskampfes in Deutschland während der Industrialisierung,* ed. Klaus Tenfelde and Heinrich Volkmann (Munich: C. H. Beck, 1981), 194–208.

10. Dirk Stegmann, *Die Erben des Bismarcks. Parteien und Verbände in der Spätphase des Wilhelminischen Deutschlands. Sammlungspolitik 1897–1918* (Cologne and Berlin: Verlag Kiepenheuer & Witsch, 1970), 124ff.

11. Dieter Lindenlaub offers a somewhat similar assessment in "Firmengeschichte und Sozialpolitik. Zur Wissenschaftsgeschichte der Nationalökonomie in Deutschland vor dem Ersten Weltkrieg," in *Wissenschaft, Wirtschaft und Technik. Studien zur Geschichte. Wilhelm Treue zum 60. Geburtstag,* ed. Karl-Heinz Mandegold (Munich: Bruckmann, 1969), 279. See also Stegmann, *Die Erben,* 125, 335.

12. Alexander Tille, *Die Berufsstandspolitik des Gewerbe- und Handelsstandes,* vol. 4, *Die politische Notwehr des Gewerbe- und Handelsstandes* (Berlin: Rosenbaum & Hart, 1910), 1–31.

13. The quote is from the *Brockhaus Lexikon* published in 1817 and cited in Rudolf Walther, "Stand, Klasse," in *Geschichtliche Grundbegriffe. Historisches Lexikon zur politisch-sozialen Sprache in Deutschland,* vol. 6, ed. Otto Brunner, Werner Conze, and Reinhart Kosselleck (Stuttgart: E. Klett, 1990), 277.

14. Tille, *Die Berufsstandspolitik,* 4:9, 8.

15. Tille, "Der Gewerbebund," *SWDWZ,* 16 August 1912, no. 33, 395.

16. "Zur Aufhebung der liberalen Gesellschaftsordnung," *SWDWZ,* 28 February 1908, no. 9, 59.

17. Alexander Tille, *Die Berufsstandspolitik,* vol. 2, *Der Geisteskampf gegen die gewerbliche Ertragswirtschaft* (Berlin: Rosenbaum & Hart, 1910), 28–29; Tille, *Die Arbeitgeberpartei und die politische Vertretung der deutschen Industrie,* no. 5, *Südwestdeutsche Flugschriften* (Saarbrücken: Kommissionsverlag der Buchhandlung C. Schmidtke, 1908): 11–12.

18. Tille, *Die Berufsstandspolitik,* 4:7.

19. See the explicit call for the reformulation of liberal doctrine on an "occupa-

tional-estatist basis" (*berufsständischer Grundlage*) in Tille, *Die Berufsstandspolitik,* 2:29.

20. Tille, "Die Ungleichheit der Menschen in ihrer Leistungsfähigkeit," *SIH,* 18 May 1906, no. 20, 125–26.

21. Tille, *Die Berufsstandspolitik,* 2:126–27, 153; Tille, "Konjunktur, Leistungsunterschied und Ständebildung," *NSGB,* 20 July 1901, no. 29, 116. See also Wilfried Schungel, *Alexander Tille (1866–1912): Leben und Ideen eines Sozialdarwinisten* (Husum: Matthiesen, 1980), 38–40, 42ff.

22. The quotes are from Tille, *Die Berufsstandspolitik,* vol. 1, *Die gewerbliche Ertragswirtschaft* (Berlin: Rosenbaum & Hart, 1910), 44ff; and "Coelibat im Volksdienst," cited in Schungel, *Alexander Tille,* 54.

23. Alexander Tille, *Vom Darwin bis Nietzsche. Ein Buch Entwicklungsethik* (Leipzig: C. G. Naumann, 1895), 208; Tille, *Der Wettbewerb weisser und gelber Arbeit in der industriellen Produktion* (Berlin: Otto Elsner, 1904). On the attitudes of industrialists more generally, see Stegmann, *Die Erben,* 282; and Fritz Fischer, *Krieg der Illusionen: Die deutsche Politik von 1911 bis 1914* (Düsseldorf: Droste Verlag, 1969), 64–77.

24. Karl Alfred Gabel, *Kämpfe und Werden der Hüttenarbeiter-Organisationen an der Saar* (Saarbrücken: Saarbrücker Druckerei und Verlag, 1921), 162.

25. Gabel, *Kämpfe und Werden,* 163. On the Saar league, see LHAK, 442/3793, 241–44; and *SZ,* 12 March 1912, no. 71, in LHAK, 442/3793, 239.

26. Klaus Mattheier, *Die Gelben. Nationale Arbeiter zwischen Wirtschaftsfrieden und Streik* (Düsseldorf: Pädagogischer Verlag Schwann, 1973), 13–24; Mattheier, "Werkvereine und wirtschaftsfriedlich-nationale (gelbe) Arbeiterbewegung im Ruhrgebiet," in *Arbeiterbewegung an Rhein und Ruhr,* ed. Jürgen Reulecke (Wuppertal: Peter Hammer Verlag, 1974), 177–79; Saul, *Staat, Industrie, Arbeiterbewegung,* 134ff.; Hans-Ulrich Wehler, *Deutsche Gesellschaftsgeschichte,* vol. 3, *Von der "Deutschen Doppelrevolution" bis zum Beginn des Ersten Weltkrieges 1849–1914* (Munich: Verlag C. H. Beck, 1995), 726–27.

27. *Rede des Herrn Langerichtsdirektors Dr. Karl Röchling, Mitglied des Abgeordnetenhauses, über Die Werkvereinsbewegung. Gehalten bei Gelegenheit der Fahnenweihe des Hüttenarbeitervereins E. V. Völklingen und des 25 jährigen Stiftungsfestes des Hüttenvereins "Stahl u. Eisen" zu Völklingen am 18. Juli 1909* (Völklingen: C. H. Scheuer, [1909], 9; letter from the Röchling management, cited in Wilhelm Kulemann, *Die Berufsvereine. Geschichtliche Entwicklung der Berufsorganisationen der Arbeitnehmer und Arbeitgeber aller Länder,* vol. 2 (Jena: G. Fischer, 1908), 466 n. 2; Gabel, *Kämpfe und Werden,* 165–66, 117; Gerhard Ames, "Arbeiten auf der Hütte. Lebenskontexte von Hüttenarbeitern," in van Dülmen, *Industriekultur an der Saar,* 118.

28. The Röchling quote is cited in Gabel, *Kämpfe und Werden,* 162.

29. Ames, "Arbeiten auf der Hütte," 118; Max Schlenker, *Streiflichter auf künftige Aufgaben wirtschaftlicher Interessenvertretungen,* no. 31, *Südwestdeutsche Flugschriften* (Saarbrücken: Kommissionsverlag der Buchhandlung C. Schmidtke, 1914), 28ff.

30. Ehrenberg, *Zur Einführung in die exakt-vergleichende Wirtschaftsforschung. Aufsätze aus dem Archiv für exakte Wirtschaftsforschung (Thünen Archive)* (Jena:

G. Fischer, 1912), 46, 67, 71–75, 79–80, 90–91, 94, 103. See also Mattheier, *Die Gelben,* 15–17, 173; and Joan Campbell, *Joy in Work, German Work: The National Debate, 1800–1945* (Princeton, NJ: Princeton University Press, 1989), 101–2.

31. Gabel, *Kämpfe und Werden,* 117.

32. "Jahresbericht des Hüttenarbeiter-Verein e.V. Geislautern, Jahrgang 1908," in LHAK, 442/3820, 306.

33. Ibid., 307.

34. See the analysis to this effect in LR (Saarbrücken) to RP, 6 December 1910, in LHAK, 442/3785, 22–23.

35. "Jahresbericht des Hüttenarbeiter-Verein e.V. Geislautern," 307, 311.

36. For the active involvement of workers in company union organizations and institutions in Burbach, Altenwald, Sulzbach, Geislautern, and Völklingen, see Gabel, *Kämpfe und Werden,* 163, 168–69; the statutes of the Röchling Hüttenarbeiterverein Koksofenanlage Altenwald, 30 August 1907, in LAS, Depositum Sulzbach, Fach 68, no. 5, vol. 1; "Jahresbericht des Hüttenarbeiter-Verein e.V. Geislautern," 302; and Gustav Beck, *50 Jahre Allgemeine Baugenossenschaft Völklingen 1904–1954* (Völklingen and Saar: Buchdruckerei Heinrich Immich, 1954), 39.

37. The quotes are taken from the Halberg, Röchling, and Burbach yellow union statutes. See Gabel, *Kämpfe und Werden,* 168; Kulemann, *Die Berufsvereine,* 464–65; and the statutes of the Röchling Hüttenarbeiterverein Koksofenanlage Altenwald, 30 August 1907, in LAS, Depositum Sulzbach, Fach 68, no. 5, vol. 1.

38. *Rede des Herrn Langerichtsdirektors Dr. Karl Röchling,* 10.

39. "Jahresbericht des Hüttenarbeiter-Verein e.V. Geislautern," 311–12.

40. Gabel, *Kämpfe und Werden,* 169; *NVZ,* 8 November 1912, no. 978, in LHAK, 442/3739, 677; Mattheier, *Die Gelben,* 354–55.

41. *SWDWZ,* 9 June 1911, no. 23, 219.

42. Gabel, *Kämpfe und Werden,* 164.

43. *SIH,* 20 October 1905, no. 42, 203–4. See also *BF,* 20 May 1905, no. 57, 459–60; *SIH,* 26 January 1906, no. 4, 25–26; the employer memorandum in *SW,* 28 April, 1906, no. 99, supplement; *SW,* 28 April 1906, no. 99, supplement; and *SW,* 30 April 1906, no. 100.

44. *SIH,* 20 October 1905, no. 42, 204.

45. *SIH,* 10 July 1903, no. 28, 137; *SIH,* 4 December 1903, no. 49, 241; SIH, 20 October 1905, no. 42, 204; and *SWDWZ,* 26 January 1906, no. 4, 25–26.

46. *SWDWZ,* 5 April 1912, no. 14, 223–34.

47. Ibid.; *Rede des Herrn Langerichtsdirektors Dr. Karl Röchling,* 2, 10.

48. Ullmann, "Unternehmerschaft, Arbeitgeberverbände und Streikbegewung," 198.

49. "Protokoll über die konstituierende Versammlung des Arbeitgeberverbandes der Brauereien Südwestpreussens, der Pfalz und benachbarter Gebiete," 27 April 1904, and "Bericht über die konstituierende Versammlung des Arbeitgeberverbandes für das Baugewerbe und die verwandten Berufe der Saargegend," 17 July 1904, in LHAK, 442/4249, 125–33, 197–203.

50. Saul, *Staat, Industrie, Arbeiterbewegung,* 360.

51. Ullmann, "Unternehmerschaft, Arbeitgeberverbände und Streikbegewung," 195, 197, 199ff.

52. Ullmann, *Interessenverbände,* 118.

53. "Verhandlungen des Vereins für Socialpolitik über das Arbeitsverhältnis in den privaten Riesenbetrieben," in *Schriften des Vereins für Socialpolitik, vol. 116, Verhandlung der Generalversammlung in Mannheim, 25., 26., 27. und 28. September 1905* (Leipzig: Duncker & Humblot, 1906), 157.

54. Cited in Stegmann, *Die Erben,* 169; Saul, *Staat, Industrie, Arbeiterbewegung,* n. 16, 186.

55. For the statements of Bueck, see *Verein Deutscher Eisen- und Stahlindustrieller, Generalversammlung,* in Bundesarchiv Koblenz, R. 131, no. 163; and Saul, *Staat, Industrie, Arbeiterbewegung,* 439 n. 87. On Hugenberg's activities, see Guratzsch, *Macht durch Organisation. Die Grundlegung des Hugenbergschen Presseimperiums* (Düsseldorf: Bertelsmann Universitätsverlag, 1974).

56. Hans-Peter Ullmann, *Der Bund der Industriellen. Organisation, Einfluss und Politik klein- und mittelbetrieblicher Industrieller im Deutschen Kaiserreich 1895–1914* (Göttingen: Vandenhoeck & Ruprecht, 1976), 103–4; Saul, *Staat, Industrie, Arbeiterbewegung,* n. 56, 437.

57. Ullmann, *Der Bund der Industriellen,* 109.

58. Cited in Stegmann, *Die Erben,* 167.

59. Saul, *Staat, Industrie, Arbeiterbewegung,* 58ff.

60. See *SIH,* 11 December 1903, no. 50, 245; *SIH,* no. 51, 18 December 1903, 254; and *SIH,* 5 January 1904, no. 2, 6. The quotes are from "Der Südwestdeutschen Wirtschaftszeitung zum Geleite," *SWDWZ,* 6 July 1906, no. 33, 183–84.

61. "Der Südwestdeutschen Wirtschaftszeitung zum Geleite," 183–84; report in *SWDWZ,* 18 May 1906, no. 20, 119.

62. Cited in Saul, *Staat, Industrie, Arbeiterbewegung,* 101.

63. Saul, *Staat, Industrie, Arbeiterbewegung,* 101, 113; Kulemann, *Die Berufsvereine.*

64. Tille, *Die politische Arbeitgeberbewegung,* no. 8, *Südwestdeutsche Flugschriften* (Saarbrücken: Kommissionsverlag der Buchhandlung C. Schmidtke, 1909), 4.

65. See the VWGWISI report in *GN,* 30 March 1904, in LHAK, 444/4244, 705.

66. Karl Erich Born, *Staat und Sozialpolitik Seit Bismarks Sturz. Ein Beitrag zur Geschichte der innenpolitischen Entwicklung des Deutschen Reiches 1890–1914* (Wiesbaden: F. Steiner, 1957), 178–242; E. P. Hennock, *The Origin of the Welfare State in England and Germany, 1850–1914: Social Policies Reconsidered* (Cambridge: Cambridge University Press), 200–211.

67. Tille, "Der Gewerbebund," 394.

68. These comments were published in the *Deutsche Industriezeitung,* cited in Alexander Tille, *Die politische Arbeitgeberbewegung,* 6.

69. Cited in Stegmann, *Die Erben,* 148.

70. The National Liberal Party's provincial leadership in Cologne, believing Tille was a member of the party, announced his ouster in late July 1906. See "Ein wirtschafliches Einlenken der nationalliberal Partei?" *SWDWZ,* 30 April 1909, no. 18, 203–4.

71. Tille, *Die Arbeitgeberpartei und die politische Vertretung,* 1, 12–13.

72. See the report of the employers' meeting in *SWDWZ,* 5 June 1908, no. 23, 145.

73. "Die wirtschaftliche Bedeutung der Reichstagswahlen," *SWDWZ,* 8 February 1907, no. 6, 46–47; "Die Selbstbesinnung des Liberalismus," *SWDWZ,* 28 February 1908, no. 9, 60; "Eine neue Absage aus Industriekreisen an die nationalliberale Partei," *SWDWZ,* 6 November 1908, no. 45, 299; "Die angebliche und die wirkliche Zusammensetzung der nationalliberalen Partei," *SWDWZ,* 8 January 1909, no. 2, 12–14.

74. Josef Bellot, *Hundert Jahre politisches Leben an der Saar unter preussischer Herrschaft (1815–1918)* (Bonn: Ludwig Röhrscheid Verlag, 1954), 230–31.

75. Stegmann, *Die Erben,* 51ff.; Tille *Die politische Arbeitgeberbewegung,* 9–24; Hans Jaeger, *Unternehmer in der deutschen Politik (1890–1918)* (Bonn: Röhrscheid, 1967), 126–28.

76. Tille, "Der Gewerbebund," 394; Tille, *Die Arbeitgeberpartei und die politische Vertretung,* 4–8.

77. Peter-Christian Witt, *Die Finanzpolitik des Deutschen Reiches von 1903–1913. Eine Studie zur Innenpolitik des Wilhelminischen Deutschlands* (Lübeck: Matthiesen, 1970); Siegfried Mielke, *Der Hansa-Bund für Gewerbe, Handel und Industrie 1909–1914* (Göttingen: Vandenhoeck & Ruprecht, 1976), 29–30, 32ff; Stegmann, *Die Erben,* 177–79.

78. Mielke, *Der Hansa-Bund.*

79. Jaeger, *Unternehmer in der deutschen Politik,* 154–55.

80. The *Aufruf* is printed in *SWDWZ,* 30 July 1909, no. 31, 321.

81. Tille, "Der Hansabund," in *Die Gründungsreden der Ortsgruppe Saarbrücken des Hansabundes, gehalten am 23. Juli 1909 im Saalbau zu Saarbrücken,* no. 10, *Südwestdeutsche Flugschriften,* ed. Tille (Saarbrücken: Kommissionsverlag der Buchhandlung C. Schmidtke, 1909), 8–10, 14–15.

82. *Die Gründungsreden,* 5–6.

83. *SWDWZ,* 30 July 1909, no. 31, 321.

84. *SWDWZ,* 12 November 1909, no. 46, 437–38.

85. Heinrich August Winkler, *Mittelstand, Demokratie und Nationalsozialismus. Die politische Entwicklung von Handwerk und Kleinhandel in der Weimarer Republik* (Cologne: Kiepenheuer & Witsch, 1972); Robert Gellately, *The Politics of Economic Despair: Shopkeepers and German Politics, 1890–1914* (London: Sage Publications, 1974); Shulamit Volkov, *The Rise of Popular Antimodernism in Germany: The Urban Master Artisans, 1873–1896* (Princeton, NJ: Princeton University Press, 1978).

86. *Die Gründungsreden,* 8, 22.

87. *Die Gründungsreden,* 43.

88. Ibid., 26–27.

89. *SWDWZ,* 25 November 1910, no. 47, 302; see the description of the course (*Lehrgang*) in *SWDWZ,* 11 September 1911, no. 36, 130.

90. See the 1910 annual report from the local branch in *SWDWZ,* 23 June 1911, no. 25, 244.

91. Tille, *Die Berufsstandspolitik,* 4:252–335. The quote is from p. 335.

92. See the annual report of the Förderungs-Ausschuss für die wirtschafts-

friedliche nationale Arbeiterbewegung for 1913–14 in LHAK, 442/3788, 653–57. On the convergence of interests between heavy industrial organizations and the yellow union movement more generally, see Dirk Stegmann, "Zwischen Repression und Manipulation. Konservative Machteliten und Arbeiter- und Angestelltenbewegung 1910–1918," *Archiv für Sozialgeschichte* 12 (1972): 354–65.

93. Cited from *SWDWZ* in Gabel, *Kämpfe und Werden,* 119–21.

94. Tille, "Der Gewerbebund," 394–95. See Stegmann, *Die Erben,* 331–35, for a discussion of heavy industry's *Mittelstandspolitik* after 1911.

95. Geoff Eley, *Reshaping the German Right: Radical Nationalism and Political Change after Bismarck* (New Haven: Yale University Press, 1980), 316ff.

96. Dirk Stegmann, "Hugenberg contra Stresemann: Die Politik der Industrieverbände am Ende des Kaiserreichs," *Vierteljahrshefte für Zeitgeschichte* 24 (1976): 351.

97. Cited in Stegmann, *Die Erben,* 362–63; Fischer, *Krieg der Illusionen,* 384–412; Saul, *Staat, Industrie, Arbeiterbewegung,* 339ff., Eley, *Reshaping the German Right,* 316–17.

98. Cited in Hartmut Kaelble, *Industrielle Interessenpolitik in der Wilhelminischen Gesellschaft. Centralverband Deutscher Industrieller 1895–1914* (Berlin: Walter de Gruyter & Co., 1967), 134.

99. For Schlenker's larger corporatist criticisms, see his *Abänderung des Reichstagswahlrechts oder Schaffung eines Reichsoberhauses?* no. 27, *Südwestdeutsche Flugschriften* (Saarbrücken: Kommissionsverlag der Buchhandlung C. Schmidtke, 1913), 14ff.; *Arbeiterschutzgesetzgebung und ihre wirtschaftlichen Rückwirkungen,* no. 30, *Südwestdeutsche Flugschriften* (Saarbrücken: Kommissionsverlag der Buchhandlung C. Schmidtke, 1914), 1–17; and *Neuregelung unserer Handelsbeziehungen: Streiflichter auf künftige Aufgaben wirtschaftlicher Interessenvertretungen,* no. 31, *Südwestdeutsche Flugschriften* (Saarbrücken: Kommissionsverlag der Buchhandlung C. Schmidtke, 1914), 15–48.

100. Stegmann, *Die Erben,* 374–77; Stegmann, "Hugenberg contra Stresemann," 350–51.

101. Stegmann, *Die Erben,* 377.

EPILOGUE

1. Mary Nolan, *Visions of Modernity: American Business and the Modernization of Germany* (New York and Oxford: Oxford University Press, 1994), 179–205; Peter Hinrichs, *Um die Seele des Arbeiters: Arbeitspsychologie, Industrie und Betriebssoziologie in Deutschland* (Cologne: Pahl-Rugenstein, 1981), 277–303; Gerhard P. Bunk, *Erziehung und Industriearbeit. Modelle betrieblichen Lernens und Arbeitens Erwachsener* (Weinheim and Basel: Belz, 1972), 204–92.

2. Hinrichs, *Um die Seele des Arbeiters,* 287.

3. Nolan, *Visions of Modernity,* 187; Hinrichs, *Um die Seele des Arbeiters,* 287. On the language of employers, see Bernd Weisbrod, *Schwerindustrie in der*

Weimarer Republic. Interessenpolitik zwischen Stabilisierung und Krise (Wuppertal: Peter Hammer Verlag, 1978), 495–501.

4. Eberhard Heuel, *Der umworbene Stand. Die ideologische Integration der Arbeiter im Nationalsozialismus 1933–1935* (Frankfurt am Main and New York: Campus Verlag, 1989), especially 321–51.

5. Tim Mason, *Sozialpolitik im Dritten Reich. Arbeiterklasse und Volksgemeinschaft* (Opladen: Westdeutscher Verlag, 1977), especially 99–123; Rüdiger Hachtmann, *Industriearbeit im "Dritten Reich." Untersuchungen zu den Lohn- und Arbeitsbedingungen in Deutschland 1933–1945* (Göttingen: Vandenhoeck & Ruprecht, 1989), 254–301; Tilla Siegel, *Leistung und Lohn in der nationalsozialistischen "Ordnung der Arbeit"* (Opladen: Westdeutscher Verlag, 1989); Heuel, *Der umworbene Stand,* 539–66; Joan Campbell, *Joy in Work, German Work: The National Debate, 1800–1945* (Princeton, NJ: Princeton University Press, 1989), 261–65, 312—-36.

6. This is detailed in Matthias Frese, *Betriebspolitik im "Dritten Reich." Deutsche Arbeitsfront, Unternehmer und Staatsbürokratie in der westdeutschen Großindustrie, 1933–1939* (Paderborn: F. Schöningh, 1991), 100–13. See also Tim Mason, "Zur Entstehung des Gesetzes zur Ordnung der nationalen Arbeit, vom 20. Januar: Ein Versuch über das Verhältnis 'archäischer' und 'moderner' Momente in der neuesten deutschen Geschichte," in *Industrielles System und politischen Entwicklung in der Weimarer Republik,* ed. Hans Mommsen, Dietmar Petzina, and Bernd Weisbrod (Düsseldorf: Droste Verlag, 1974), 322–51.

7. Frese, *Betriebspolitik im "Dritten Reich,"* 341–50.

8. Martin Geyer, "Soziale Sicherheit und wirtschafltlicher Fortschritt. Überlegungen zum Verhältnis von Arbeitsideologie und Sozialpolitik im 'Dritten Reich.'" *Geschichte und Gesellschaft* 15 (1989): 382–406; Hachtmann, *Industriearbeit im "Dritten Reich,"* 83–89, 237–44; Robert Proctor, *Racial Hygiene: Medicine under the Nazis* (Cambridge, MA: Harvard University Press, 1988), 269–70.

9. Neil Gregor, *Daimler-Benz in the Third Reich* (New Haven: Yale University Press, 1998), 165–69; Carola Sachse, *Siemens, der Nationalsozialismus und die moderne Familie. Eine Untersuchung zur sozialen Rationalisierung in Deutschland im 20. Jahrhundert* (Hamburg: Rasch und Röhring Verlag, 1990), 42–53; Hans Mommsen and Manfred Grieger, *Das Volkswagenwerk und seine Arbeiter im Dritten Reich* (Düsseldorf: ECON, 1996), 227–49, especially 232.

10. Ulrich Herbert, *Fremdarbeiter. Politik und Praxis des "Ausländer-Einsatzes" in der Kriegswirtschaft des Dritten Reiches* (Berlin and Bonn: Verlag J. H. W. Dietz Nachf., 1985).

11. Heinz Hagenlücke, *Deutsche Vaterlandspartei. Die nationale Rechte am Ende des Kaiserreiches* (Düsseldorf: Droste Verlag, 1997).

12. Dirk Stegmann, *Die Erben Bismarcks. Parteien und Verbände in der Spätphase des Wilhelminischen Deutschlands. Sammlungspolitik 1897–1918* (Cologne: Kiepenheuer & Witsch, 1970), 511–13.

13. Weisbrod, *Schwerindustrie,* 154–55.

14. Christa Thieme, "Deutsche Industriellenvereinigung (DI) 1924–1934," in *Die bürgerlichen Parteien in Deutschland. Handbuch der Geschichte der bürgerlichen Parteien und anderer bürgerlicher Interessenorganisationen vom Vormärz biz sum*

Jahre 1945, vol. 1, ed. Dieter Fricke (Leipzig: Verlag Enzyklopädie Leipzig, 1968), 387–89.

15. Charles S. Maier, *Recasting Bourgeois Europe: Stabilization in France, Germany, and Italy in the Decade after World War I* (Princeton, NJ: Princeton University Press, 1975).

16. Kevin Passmore, *From Liberalism to Fascism: The Right in a French Province, 1928–1939* (Cambridge: Cambridge University Press, 1997), 167–85, especially 185; Michael Richards, *A Time of Silence: Civil War and the Culture of Repression in Franco's Spain, 1936–1945* (Cambridge: Cambridge University Press, 1998), 110–26, especially 121–22; Maier, *Recasting Bourgeois Europe,* 560–78.

17. Ralph H. Bowen, *German Theories of the Corporative State* (New York: Whittlesey House, 1947), 10.

18. Avraham Barkai, *Das Wirtschaftssystem des Nationalsozialismus. Ideologie, Theorie, Politik 1933–1945* (Frankfurt am Main: Fischer Taschenbuch Verlag, 1988), 32.

19. Dirk Stegmann, "Zum Verhältnis von Großindustrie und Nationalsozialismus 1930–1933. Ein Beitrag zur Geschichte der sog. Machtergreifung," *Archiv für Sozialgeschichte* 23 (1973): 421–22; Michael Schneider, *Unternehmer und Demokratie. Die freien Gewerkschaften in der unternehmerischen Ideologie der Jahre 1918 bis 1933* (Bonn and Bad Godesberg: Verlag Neue Gesellschaft, 1975), 181; Reinhard Neebe, *Großindustrie, Staat und NSDAP 1930–1933. Paul Silverberg und der Reichsverband der Deutschen Industrie in der Krise der Weimarer Republik* (Göttingen: Vandenhoeck & Ruprecht), 117.

20. Heinrich August Winkler, "Unternehmerverbände zwischen Ständeideologie und Nationalsozialismus" *Viertelsjahrshefte für Zeitgeschichte* 17 (1969): 341–71; Claus-Dieter Krohn, "Autoritärer Kapitalismus. Wirtschaftskonzeptionen im Übergang von der Weimarer Republik zum Nationalsozialismus," in *Industrielle Gesellschaft und politisches System. Beiträge zur politischen Sozialgeschichte,* ed. Dirk Stegmann, Bernd-Jürgen Wendt, and Peter-Christian Witt (Bonn: Verlag Neue Gesellschaft, 1978), 120ff.; Stegmann, *Die Erben,* 520–21.

21. Stegmann, "Zum Verhältnis von Großindustrie und Nationalsozialismus," 452–65. The quote is from p. 463.

22. Stegman, "Zum Verhältnis von Großindustrie und Nationalsozialismus," especially 430; Neebe, *Großindustrie,* 132–34; David Abraham, *The Collapse of the Weimar Republic: Political Economy and Crisis,* 2nd ed. (New York: Homes & Meier, 1986), 310–12; Dick Geary, "The Industrial Elite and the Nazis in the Weimar Republic," in *The Nazi Machtergreifung,* ed. Peter D. Stachura (London: George Allen & Unwin, 1983), 85–100, Winkler, "Unternehmerverbände zwischen Ständeideologie und Nationalsozialismus," 361–62. See also Barkai, *Das Wirtschaftssystem,* 46–47. For a contrary view, see Henry Turner, *German Big Business and the Rise of Hitler* (New York: Oxford University Press, 1985), 272–339.

23. Heuel, *Der umworbene Stand,* 321–29, 341–51; Schneider, *Unternehmer und Demokratie,* 180–82.

24. It is important to note that even leading Nazis differed over their corporatist commitments. See the evidence in Winkler's "Unternehmerverbände

zwischen Ständeideologie und Nationalsozialismus" (355–57), which nevertheless assumes a single vision.

25. Peter Hayes, "Polycracy and Policy in the Third Reich: The Case of the Economy," in *Reevaluating the Third Reich,* ed. Thomas Childers and Jane Caplan (New York: Holmes & Meier, 1993), 199–200.

26. See Neebe, *Großindustrie,* 184–88 (despite Neebe's contrary argument); and Arthur Schweitzer, *Big Business in the Third Reich* (London: Eyre & Spottiswoode, 1964), 124–28. The plan for compulsory cartels that was finally adopted was prepared by a committee within the RDI.

27. Tooze, *Wages of Destruction: The Making and Breaking of the Nazi Economy* (London: Penguin Books, 2006), 562–64. See also Hans-Erich Volkmann, *Ökonomie und Expansion. Grundzüge der NS-Wirtschaftspolitik. Ausgewählte Schriften* (Munich: R. Oldenbourg Verlag, 2003), 76.

28. Dietmar Petzina, *Autarkiepolitik im Dritten Reich. Der nationalsozialistische Vierjahresplan.* (Stuttgart: Deutsche Verlags-Anstalt, 1968), 197; Barkai, *Das Wirtschaftssystem,* 16–17, 98.

29. Michel Foucault, "Governmentality," in *The Foucault Effect: Studies in Governmentality,* ed. Graham Burchell, Colin Gordon, and Peter Miller (Chicago: University of Chicago Press, 1991), 103.

30. I have here drawn on and altered Bob Jessop's definition of Fordism in his essay "Fordism and the State," in *Post-Fordism: A Reader,* ed. Ash Amin (London: Blackwell, 1994), 252–54.

31. An obvious example in this regard would be the influential and allegedly "moderate" (i.e., non-Nazi) Ruhr industrialist Paul Reusch, chairman of the Langnamverein, who expressed tactical opposition to a Hitler-led government in early January 1933 but had allied himself with Hitler and the Nazi movement during the fall of 1931 in the run-up to the presidential elections in the summer of 1932, before turning back to a Papen-led governing coalition, with Nazi participation, during the fall and winter of 1932. It was Franz von Papen, of course, who finally orchestrated Hitler's appointment as chancellor in January 1933. Neebe, *Großindustrie,* 119–22, 131–32, 144–45; Abraham, *Collapse of the Weimar Republic,* 161–63, 310–11.

32. Hagenlücke, *Deutsche Vaterlandspartei,* 190.

33. Nolan, *Visions of Modernity,* 188.

34. Weisbrod, *Schwerindustrie,* 243–44; Abraham, *Collapse of the Weimar Republic,* 127; Neebe, *Großindustrie,* 44–49.

35. Weisbrod, *Schwerindustrie,* 441 n. 173, 446.

36. Neebe, *Großindustrie,* 54.

37. Ibid., 60.

38. Cited in Winkler, "Unternehmerverbände zwischen Ständeideologie und Nationalsozialismus," 351.

39. This formulation is paraphrased by Weisbrod in *Schwerindustrie,* 495–96.

40. Abraham, *Collapse of the Weimar Republic,* 305.

41. Eberhard Czichon, *Wer verhalf Hitler zur Macht? Zum Anteil der deutschen Industrie an der Zerstörung der Weimarer Republik* (Cologne: Pahl-Rugenstein, 1967), 23, 67. At the time, the leader of this coalition-ready wing was Gregor

Strasser. His support from Ruhr heavy industry is detailed in Neebe, *Großindustrie,* 145; and Peter D. Stachura, *Gregor Strasser and the Rise of Nazism* (London: George Allen & Unwin, 1983), 93–94.

42. Stegmann, "Zum Verhältnis von Großindustrie und Nationalsozialismus," 468–75.

43. Cited in Winkler, "Unternehmerverbände zwischen Ständeideologie und Nationalsozialismus," 351–52.

44. Neebe, *Großindustrie,* 187–88.

45. Stegmann, *Die Erben,* 340, 508.

46. Fritz Jacoby, *Die nationalsozialistische Herrschaftsübernahme an der Saar. Die innenpolitischen Probleme der Rückgliederung des Saargebietes bis 1935* (Saarbrücken: Minerva-Verlag Thinnes & Nolte, 1973), 35–37.

47. This is inferred from Bunk, *Erziehung und Industriearbeit,* 112–90; and Hinrichs, *Um die Seele des Arbeiters,* 293–94.

48. Kurt Gossweiler, "Bund zur Erneuerung des Reiches (BER) 1928–1933," in *Die bürgerlichen Parteien in Deutschland. Handbuch der Geschichte der bürgerlichen Parteien und anderer bürgerlicher Interessenorganisationen vom Vormärz bis zum Jahre 1945,* vol. 1, ed. Dieter Fricke (Berlin: Das Europäische Buch, 1968), 195–200; Gerald D. Feldman, "The Social and Economic Policies of German Big Business, 1918–1929," *American Historical Review* 75 (1969): 53.

49. Gerhard Paul, *Die NSDAP des Saargebietes 1920–1935. Der verspätete Aufstieg der NSDAP in der katholischen-proletarischen Provinz* (Saarbrücken: Druckerei und Verlag, 1987), 246.

50. Jacoby, *Die nationalsozialistische Herrschaftsübernahme an der Saar,* 103ff.; Paul, *Die NSDAP des Saargebietes,* 62, 73–74; Gerhard Paul, *"Deutsche Mutter—heim zu Dir!" Warum es mißlang, Hitler an der Saar zu schlagen. Der Saarkampf 1933 bis 1935* (Cologne: Bund-Verlag, 1984), 69–77.

51. Paul, *"Deutsche Mutter,"* 132–49, 221.

52. Cited in ibid., 217–18.

53. Herrmann Röchling, *Wir halten die Saar!* (Berlin: Volk und Reich Verlag, 1934), 133.

54. He became a party member on 1 November 1935. See Paul, *Die NSDAP des Saargebietes,* 77.

55. Klaus-Michael Mallmann and Gerhard Paul, *Herrschaft und Alltag. Ein Industrierevier im Dritten Reich. Widerstand und Verweigerung im Saarland 1935–1945,* vol. 2 (Bonn: Dietz, 1991), 49, 145–63; Paul, *Die NSDAP des Saargebietes,* 129, 160.

56. Mallmann and Paul, *Herrschaft und Alltag,* 314.

57. Gabriele Lotfi, *KZ der Gestapo. Arbeitserziehungslager im Dritten Reich* (Frankfurt am Main: Fischer Taschenbuch Verlag, 2003), 58ff. On the terror directed at industrial workers generally, see Tim Mason, *Social Policy in the Third Reich: The Working Class and the "National Community"* (Providence and Oxford: Berg, 1993), 331–69.

58. Mallmann and Paul, *Herrschaft und Alltag,* 308–10.

59. Bernhard R. Kroener, Rolf-Dieter Müller, and Hans Umbreit, *Das*

deutsche Reich und der Zweite Weltkrieg, vol. 5, *Organisation und Mobilisierung des deutschen Machtbereichs. Zweiterhalbband, Kriegsverwaltung. Wirtshaft und personelle Ressourcen 1942–1944/45* (Stuttgart: Deutsche Verlags-Anstalt, 1999), 394.

60. See the excerpt from Röchling's 1936 *Denkschrift* to Adolf Hitler entitled "Gedanken über die Vorbereitung zum Kriege und seine Durchführung," in *Europastrategien des deutschen Kapitals 1900–1945,* ed. Reinhard Opitz (Cologne: Pahl-Rugenstein Verlag, 1977), 626–27.

61. Franz Neumann, *Behemoth: The Structure and Practice of National Socialism, 1933–1944* (New York: Harper & Row, 1966), 601. For Röchling's explicit references to the "private economy" in this context, see Tooze, *Wages of Destruction,* 571.

62. Cited in Mallmann and Paul, *Herrschaft und Alltag,* 316.

63. Kroener, Müller, and Umbreit, *Das deutsche Reich und der Zweite Weltkrieg,* 322ff.; Konrad Fuchs, "Hermann Röchling," in *Saarländische Lebensbilder,* vol. 2, ed. Peter Neumann (Saarbrücken: Saarbrücker Druckerei und Verlag, 1984), 243–46. For Röchling's explicit call in 1914 for the annexation of these regions, see his *Denkschrift* to von Dallwitz, *Statthalter* of Alsace-Lorraine, reprinted in Opitz, *Europastrategien des deutschen Kapitals,* 211–12.

64. Fuchs, "Hermann Röchling," 247–48.

65. Mitchell Dean, *Critical and Effective Histories: Foucault's Methods and Historical Sociology* (London: Routledge, 1994), 35.

Index